Semiconductor Equipments

Concepts and Applications for R & D

Semiconductor Equipments

Concepts and Applications for R & D

Manabu Tsujimura
Managing Executive Officer of Ebara Corporation
Semiconductor Equipment Division of Precision Machinery Company
Visiting Professors of Clarkson University, Tokyo Metropolitan University, Hanyang UniversityFellow of JSME (The Japan Society of Mechanical Engineering)

Comebook

Comebook,Co,134-17 Seongsan-Dong, Mapo-Gu, Seoul,121-250
Korea
Tel(82)2 337 5951
Fax(82)2 375 6075
comebook.com

Semiconductor Equipments Concepts and Applications for R&D

ISBN:978-89-957402-2-4

Printed in the Republic of Korea

Preface

My first book was published on December 3, 2007.

The book, entitled *Zukai: Handotai Wet Process Saizensen - Mekki, CMP, senjo, soshite device eno oyo* (English version: *Semiconductor Wet Process Revolution - Device Application of Plating, CMP, Cleaning*), was published by Kogyo Chosakai Publishing, Inc. It was rewritten for publication based on the textbook used at Tokyo Metropolitan University (Japan), Clarkson University (U.S.), and Hanyang University (Korea). Thankfully, the book sold well and ranked No. 18 in the "Weekly Best Sellers" of a famous bookshop.

After the first book, I became *addicted* to publishing books.

In the company I work for, so-called technical notes are released twice each month. I have already released more than 400 technical notes. By combining these technical notes and the lecture scripts I used as a visiting professor, I have decided to publish my second book, *Handotai-seizo-sochi Kaihatsu no Gokui to Jissen - Kaihatsugata-kigyo de ikinuku tebikisho - JOY (Joshiki O Yaburo)* (*Semiconductor Equipment: Secrets and Practices of Development, Manual for Surviving in a Development-based Company - JOY (Defy Common Wisdom)*). It would be great if this book were read as a companion piece to *Handotai Wet Process Saizensen* (*Wet Process Revolution*).

As an employee, I wrote technical notes first for my fellow engineers. Some of them can be made open to the public, but some cannot. Next, I wrote a textbook for university students with the intention of making every effort to make the contents of the technical notes as open as possible to the public. Since I have also been a lecturer at SEMI High Tech University, I reedited the textbook for high school students. Then, I had my 82-year old mother read the book and asked if it interested her, as I had done for the first book, and I made some revisions based on her feedback. At last, an *aweless* technical book targeting a wide audience ranging from top company engineers and university/high school students to 82-year old homemaker was born. I hope that readers will find it easy to follow.

In this book, "Chapter I: Five Rules," which describes the five development rules holding sway over Ebara Corporation (I think), and "Chapter VI: Five Revolutionary Technologies," which describes technologies practiced through the course of my development over the last 30 years, are the most spectacular. I chose the long title envisioning various sorts of readers: *Handotai-seizo-sochi Kaihatsu no Gokui to Jissen* (*Semiconductor Equipment: Secrets and Practices of*

Development) for engineers, *Kaihatsugata-kigyo de ikinuku tebikisho* (*Manual for Surviving in a Development-based Company*) for pre-career university students or entry-level employees, and *JOY (Joshiki O Yaburo)* (*Defy Common Wisdom*) for high school students and my mother. I hope that readers like it.

I believe that science and technology have evolved by **99 % imitation, 1 % innovation (Something New)**. Mere imitation is regarded as an act of a lazy mind, but I think that imitation is a good start if we consider that the efforts of our predecessors have born the fruits of evolution, innovation, and revolution. After a period of imitation ends, we should take pride as an engineer in working hard to develop new technology. Although 99 % of this book is based on the findings of our predecessors, I believe that it contains 1 % of Something New. Comments on this book from readers are always welcome.

Upon publishing this book, I would like to sincerely thank Prof. Ota of Tokyo Metropolitan University, my supervisor for my doctorate earned in 2002, Prof. Babu of Clarkson University, the first person who invited me as a visiting professor, and Prof. Park of Hanyang University, the organizer of the Japan-Korea internship program, who acts as a bridge between Japan and Korea, for their continued support. As I repeatedly mention, my dream is to establish a fund (from my modest retirement allowance) for holding a conference entitled *The Evolution of the Wet Revolution* and to invite foreign students once a year after working as a visiting professor in five nations and retiring from Ebara Corporation. I me hope that the professors help to realize my dream (what nerve I have to ask such a favor...).

I attribute the *readability* of this book (I am one of those who think this book a good read) to my students. Various questions were raised by various students. All of their questions appear in this book as training material offered through me as a digestive system. I would like to thank every one of my students.

Thanks to all who helped me write this book, particularly those who helped me acquire engineering skills, chipset/device manufacturers, material manufacturers, investment analysts, and my colleagues working hard around-the-clock.

I would also like to thank the publisher. This book never could have been published without their hard work.

Lastly, I must not forget to thank my beloved wife, Kumiko, who has been patient and helpful (she also appears somewhere in this book).

October 1, 2008
M. Tsujimur

- Contents -

Chapter II: JOY (Defy Common Wisdom)

Chapter III: How to Survive Using R&D Concepts

Chapter I

Five Rules

1. Pursue Something New.
2. Develop *must* technologies, not better technologies.
3. Think development as a cost that must be recovered.
4. Start with possessed technologies (use your expertise).
5. Think and think! Think until you feel sick!

You can do development!

You can do development! (Figure 1)

There may be an objection that "it is irresponsible to say such a thing without knowing who *you* are," but my statement is true. Just believe me and read through this book. Then, *you* will be motivated and think, "I will develop something" or "I will be the world's No. 1." Some may realistically think, "I will take patents for money," and others may say, "I will take a PhD and become a professor."

Development like groping in the dark (Figure 2)

Development is like groping in the dark.

My statement that everyone can do development only means that *everyone has a chance to do so*. Nothing can be obtained without efforts. It is advisable to find better approaches by learning from predecessors, rather than wasting efforts. To keep the most important things in mind, development and organizational doctrines will be provided first. Then, five rules for observing the doctrines will be given. I am sorry for citing a serious word *doctrine*, but development is tough work. Marketing may fail regardless of technical considerations, while there may be no right answer in engineering. Indeed, engineers have to move on in the dark, and withdrawal rules may be required. Not everyone always follows leaders; in turn, leaders may lead everyone into hell.

Leaders are lonely.

In such cases, the development and organizational doctrines and the five rules are reliable. My engineering experience in a career life of over 30 years has naturally created the two doctrines and five rules.

Looking at the forest of technology development (Figure 3)

Let's look at the forest of technology development you will go into. What do you see? There are milestones all around the forest, and the engineering gods secretly live there. I hope that you will go the right way and see the gods in your own engineering career.

Fig.1 You can do development!

Fig. 2 Development like groping in the dark

Fig.3 Looking at the forest of technology development

Development doctrine

Development doctrine (Figure 1)

The first doctrine is described below.

Business activities entail the proper cycle of Strategies (S), Research (R), Development (D), Manufacturing (M), and Sales (S). This entire cycle represents a development process in a broad sense.

For development, it is essential to create a concept by collecting market and technical information as much as possible and discussing it. Soon after determining the concept, a coupon test (simple test) is performed to make a process of reference (POR). The coupon test is important since it influences the entire subsequent course of development. Then, an in-house test version (α), user evaluation version (β), and production model (γ) are fabricated in order.

Anytime you lose the way, go back to the first step and think again. Doing so provides the proper cycle of SRDMS. Once this cycle is established, any trial, tribulation, or difficulty can be dealt with. Let's keep going with confidence.

M&A vs. in-house development (Figure 2)

It is often said that technologies are quickly acquired through M&A in Europe and the U.S., while in-house development is dominant in Japan. An increased number of Japanese companies have recently changed to apply the concept of Open Innovation, but in-house development based on NIH (Not Invented Here) (an idea that development should be made in-house) is still the mainstream in Japan.

Let's distinguish between M&A-based and in-house developments.

For in-house development, development leaders are required to have the skill of judging the development concept. Once a development project starts, it involves costs and may be difficult to cancel.

For M&A-based development, funds are typically available from organizations called Angel or venture capitals. Using these funds, venture companies carry out development work based on various information and create an α version. Judgment on the near-completed product is important.

M&A-based and in-house developments have different merits and demerits. It is recommended to select a method suitable for each company.

Fig.1 Development doctrine and best cycle of SRDMS

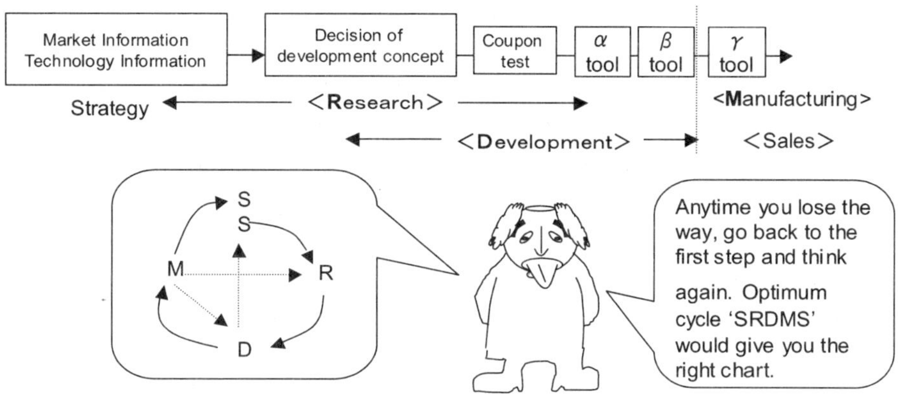

Fig.2 M&A type versus 'invented here' type

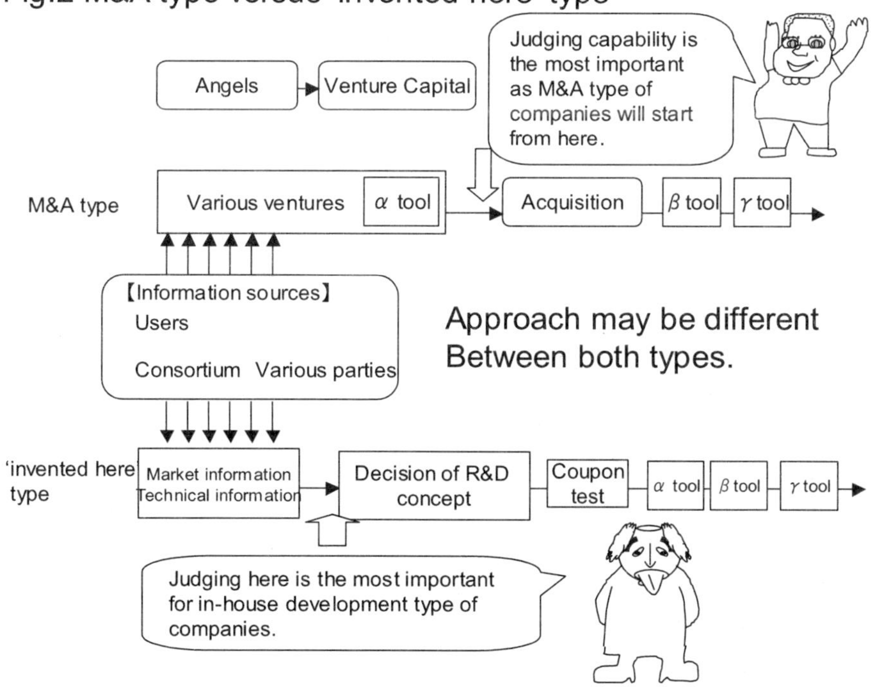

Organizational doctrine

Mission & Commitment (Figure 1)

Development projects succeed with people and their organization.

An example of organizational operation is presented below.

Figure 1 illustrates a typical business division. It has a hierarchy consisting of a division manager, managers, section chiefs, and subordinates.

The figure shows how each mission is broken down into individual levels and how commitments are made at individual levels in this division.

The division manager's mission (target to be achieved) is linked to his/her commitment to the president (promise to achieve the mission). The mission is handed down to the managers, who share commitment to the division manager. Then, the commitment to the mission is shared by the section chiefs.

Thus, a mission and commitment are combined and conveyed to lower levels.

The framework of Mission & Commitment in an organization is established in this way.

The mission given by the president is important since it influences all missions and commitments at lower levels. This framework offers various advantages, such as strengthening the sense of unity in an organization and preventing duplicate reporting.

Escalating everything? (Figure 2)

Most activities in a company are classified as routine work in which fixed or scheduled jobs are performed. At the same time, non-routine issues often occur. An organization with the framework of Mission & Commitment is effective in communicating both routine and non-routine issues.

In the Mission & Commitment framework, the expected role and responsibility of each person are specified. For an issue regarded as within a person's scope of responsibility, the person may consider it as routine and need not report it to his/her superior. A rule for immediate escalation (communication to a higher level) of any issue outside the scope of responsibility is also defined. Escalating everything is not needed. It may be appropriate that about 80 % of issues are considered as routine, and the rest (20 %) as those to be escalated.

This framework is military-like or bureaucratic (in a good sense) [1)], but an organization must have these rules to achieve successful development.

Fig.1 Mission & Commit

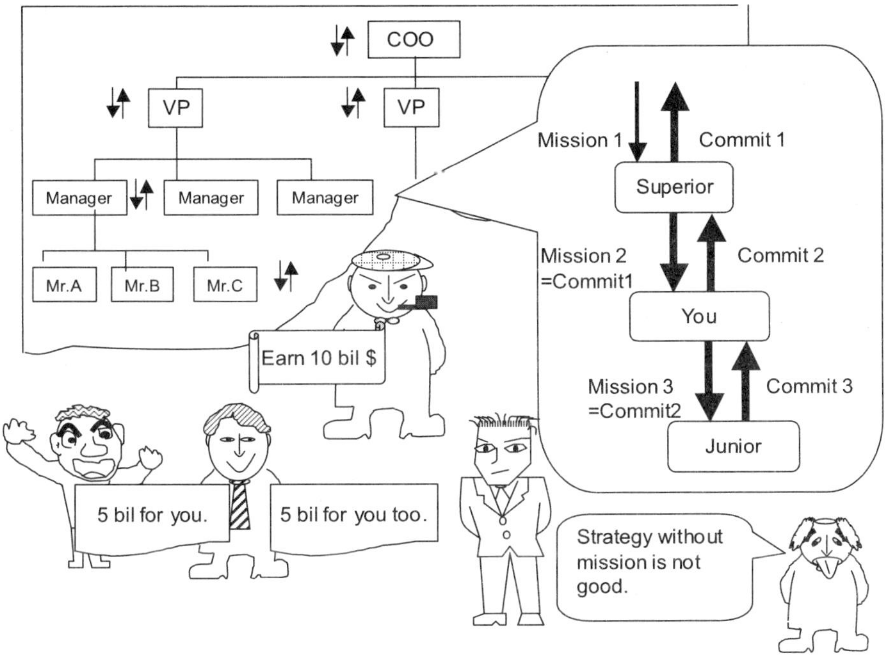

Fig. 2 Routine & non-routine

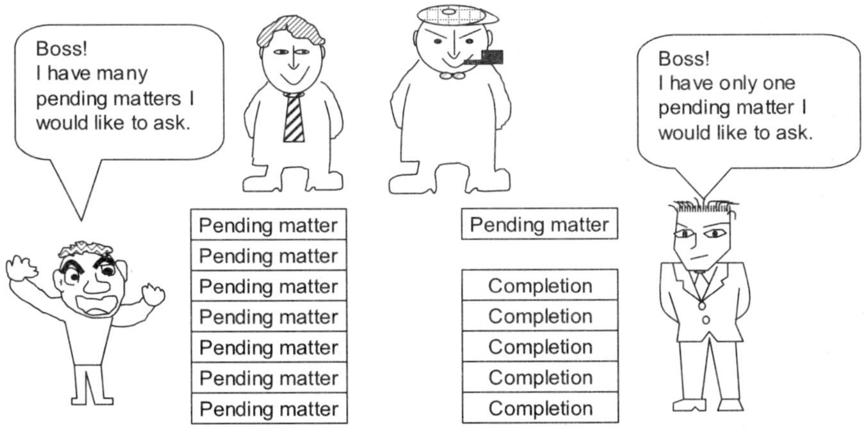

Five rules

Finally, this section provides development-related slogans.

Development requires lonely and hard work. As described earlier, engineers may grope in the dark, worrying about where the right answer is. We often misbelieve that we are approaching the answer and are soon disappointed; in such cases, I cannot help but ask the gods for help. The slogans below (five rules) will help you in such a helpless situation. I have made a point of seeing the slogans attached to the back of my company ID card when I face difficulties. Surprisingly, the slogans always give me a good idea. They are worth trying.

In fact, the slogans are (beneficial) words that cover the aspects of research, development, and manufacturing.

(1) Pursue Something New.

A philosophy that the goal (of your company) is to be a development-based company.

(2) Develop must technologies, not better technologies.

A principle for competitive strategies. Better technologies refer to conventional methods for improving operational efficiency. The goal is to develop *must* products differentiated from and having a competitive advantage over others. You must be prepared to trade off other products.

(3) Think development as a cost that must be recovered.

Profitability (ROA, ROE, and ROI). It is necessary to manage research on a cost center basis and development on a profit center basis with a management perspective. Profitability must be considered.

(4) Start with possessed technologies (use your expertise).

A warning on reckless diversification. Japanese companies tend to promote diversification and globalization by using in-house technologies since they rarely push development through M&A. They may buy necessary patents or collaborate with other companies and organizations; in this process, the companies should establish clustering strategies for development and business. This approach may be an ideal way for the Japanese version of Open Innovation (without relying on M&A).

(5) Think and think! Think until you feel sick!

A mental attitude that should be experienced at least once in achieving the goal of being a development-based company.

Fig.1 Slogans

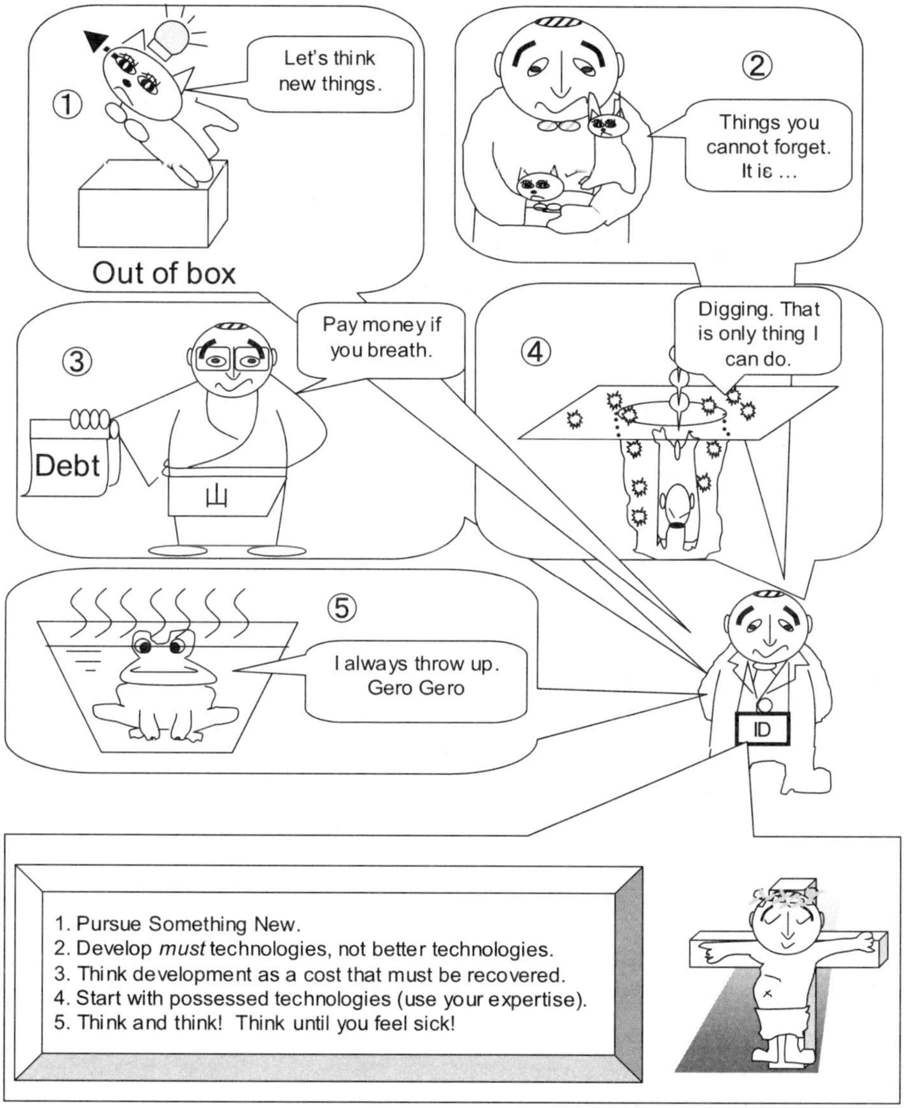

Something New for development

"Pursue Something New."

The philosophy that a company's goal is to be a development-based company is explained below.

At a development meeting (Figure 1)

One spring afternoon, a development meeting was held.
It was warm enough to induce drowsiness.
"The planarization performance of this planarization system is..." A young engineer's presentation went on.
I asked, "Is there Something New? Can you distinguish unprecedented novel features from others by using different colors in the document?"
The presenter was lost for words.
This is what I experienced.
It is important that a presenter distinguishes between what is already present in the world and what has been newly devised by the presenter.
I believe that development requires 99 % imitation and 1 % Something New.
Even 1 % Something New is sufficient.

Something New for development (Figure 2)

Following the answer to the question about Something New,
I further ask, "Is that true?"
"What makes you say so?" and
"What is the mechanism?"
Asking questions like a child asking many questions to the grandmother during a walk is essential for development.

Drawing a picture (Figure 3)

Doubtlessly, your questions will upset your subordinates. Then, why not be more unwelcome?
When asking the questions above ((1) Is it true? (2) Why? (3) What is the mechanism?), have the presenter draw a picture on a writing board. The ability to illustrate an idea is critical for development activities.
"The change in Parameter A is obviously larger than that in B." Is that really so obvious?

Fig. 1 What's new? Distinguish by color what is your idea.

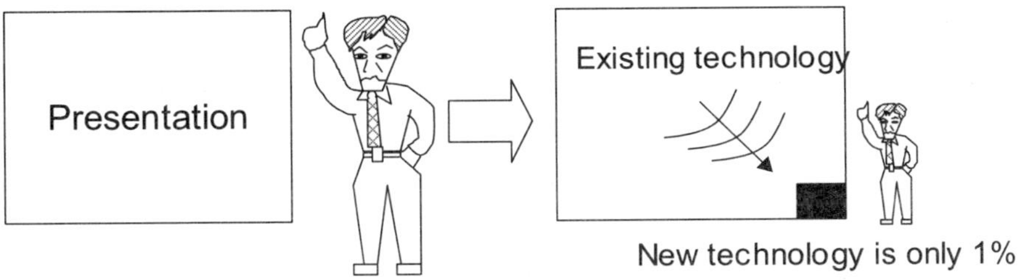

R&D is 99 % imitation and 1% something new

Fig. 2 Really? Why? Mechanism?

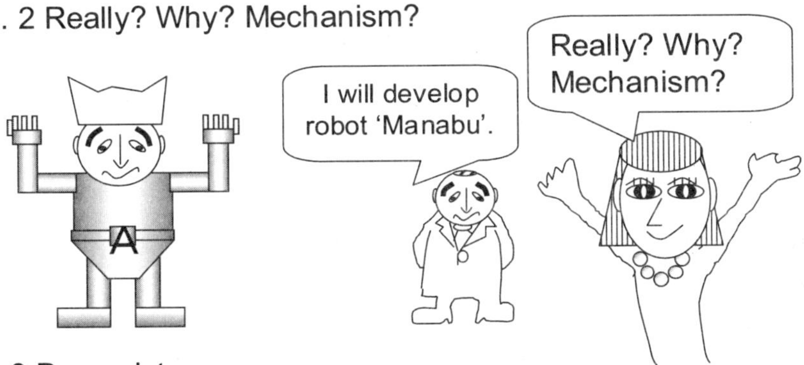

Fig. 3 Draw pictures

Clean vacuum system

Starting with oil-sealed rotary pumps (Figure 1)

It was in 1985.

In those days, oil-sealed rotary pumps were commonly used for semiconductor equipment.

If you have experience with oil-sealed rotary pumps, you may well know how troublesome they are. They store oil inside their casing and stir it with rotating vanes. They have a simple structure and provide high performance, but are sodden with oil.

The use of oil was common in those days.

Defying the common wisdom: Oil-free vacuum pump (Figure 2)

A vacuum pump using no oil was developed.

However, the pump caused a serious problem!

No oil leakage was observed downstream of the vacuum pump, but a large amount of reaction product came out.

In fact, the oil had served to trap reaction product, as well as for vacuum sealing.

Development for gas treatment (Figure 3)

The development of a system for reaction product treatment was worked on.

A dry pump was adopted as an oil-free vacuum pump.

A dry gas trap was also adopted as a dry (oil- or moisture-free) gas treatment system.

These two dry methods were combined to provide a *clean vacuum system.*

In those days, dry beer became popular, and the word *dry* boomed.

Though not for this reason, the dry system enjoyed a good reputation like dry beer.

The common wisdom was overturned to develop an oil-free vacuum pump.

Then, another problem occurred.

The process of solving the problem led to the development of a new system.

In this way, a development scheme was established.

Since then, I have been conscious of what common wisdom is and what happens if the common wisdom is overturned. This idea provides a basis for JOY as explained later. JOY: Joshiki O Yaburo (defy common wisdom)

Fig.1 in the past

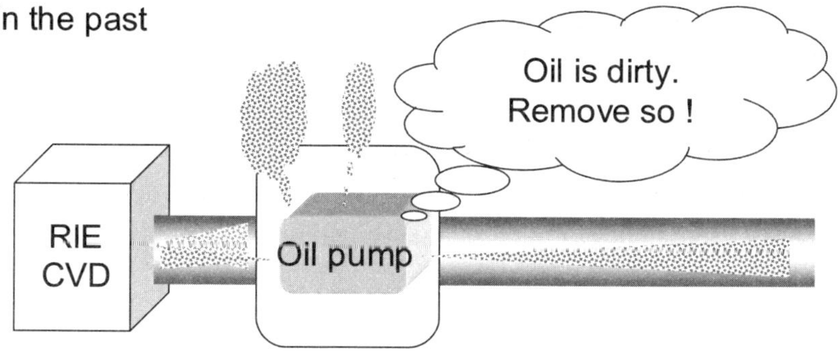

Fig. 2 Oil-free

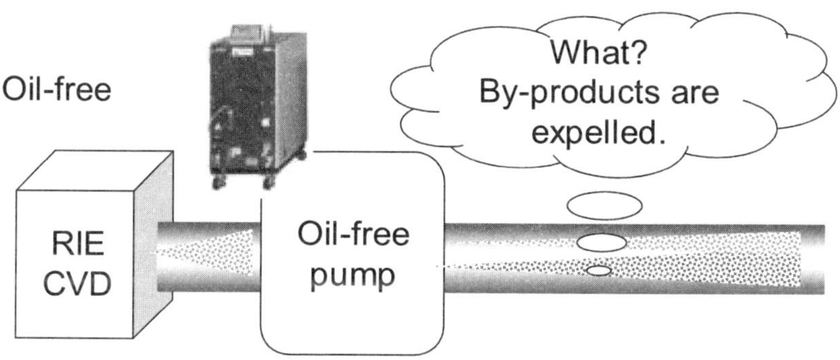

Fig. 3 Clean vacuum system

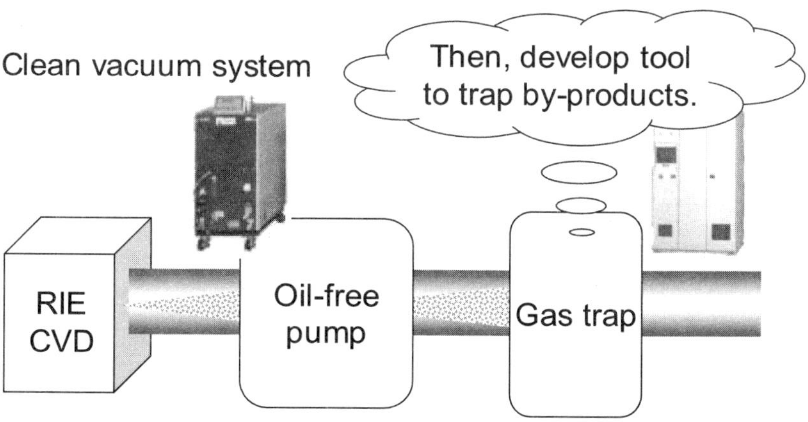

Now what do you do?

The first rule, "Pursue Something New," is as explained earlier.
Accordingly, I have owned more than 100 patents and written more than 100 papers so far.
As described later, I think that "patents benefit companies, and papers benefit society." Writing good patents facilitates writing papers.
This rule is applicable to all fields, not limited to research fields.
Here are questions. Please ask and answer them yourself.

What are you currently **researching**?
How new is it?
Is there Something New?

Can you add new functions to the product you are **designing**?
How new is it?
Is there Something New?

Can you enhance the reliability of the component you are **manufacturing**?
How new is it?
Is there Something New?

Can you change the sales method for the product you are **selling**?
How new is it?
Is there Something New?

Is there a new finding in the three **financial statements** you are preparing?
How new is it?
Is there Something New?

Do you feel that you are stuck at work?
Try to find Something New **in routine work**.
Is there Something New?

R&D
Make it clear what's new although you may think that your idea is full of new things.
Design
I reviewed existing drawings. I can make parts stronger.
Production
I can make parts lighter.
Finance
If you look more carefully PL/BS/CF, you can understand your company more.
All your business Even in your life?
Memo
Boss! It is easier to control your schedule, if I use cloud computer

CMP as only one technology

"Develop must technologies, not better technologies."

This rule expresses a principle for competitive business strategies. Better technologies refer to conventional methods for improving operational efficiency. The goal is to develop *must* products differentiated from and having a competitive advantage over others, i.e. to be the only one. You must be prepared to trade off other products.

CMP with superiority over other planarization technologies(Figure 1)

In the 1980s, CMP technology emerged and achieved unprecedented planarization performance [2)]. Figure 1 illustrates the effectiveness of CMP. The left figure shows a cross-sectional view of devices without CMP applied. The layers are non-planar. The right figure schematically shows devices with CMP applied. Since each layer is sufficiently planar, no problem is observed for the upper layers. An engineer who used CMP said a terrible thing: "CMP is like a drug. Once you use it, you can never get away from it."
CMP is the only one technology, not a better technology.

3K technology (kitanai, kitsui, and keiken-izon) (Figure 2)

CMP was not accepted from the beginning. It is based on polishing, which was notorious as so-called 3K technology (kitanai (dirty), kitsui (tough), and keiken-izon (experience-dependent)). No one could like it.

Also, wafers are wet after polishing. CMP cannot be used in clean rooms unless cleaning/drying is available. Polishing is the world's dirtiest process, while cleaning is the world's cleanest one; the dirtiest and cleanest processes must be integrated. Without this idea, everyone was reluctant to adopt CMP on the assumption that polishing could not be used for semiconductor processing.

Things turned around.

As described above, CMP offered excellent planarization performance. A desire to use it by any means came first. Desire is the father of revolutionary technologies. Some engineers tried to apply CMP and created a system based on the dry-in/dry-out concept. Now, CMP is commonly used in clean rooms.

CMP is the only one technology generated under the philosophy of "developing must technologies, not better technologies."

Fig.1 Comparison of planarization with and without CMP

Fig. 2 How to use dirty tools

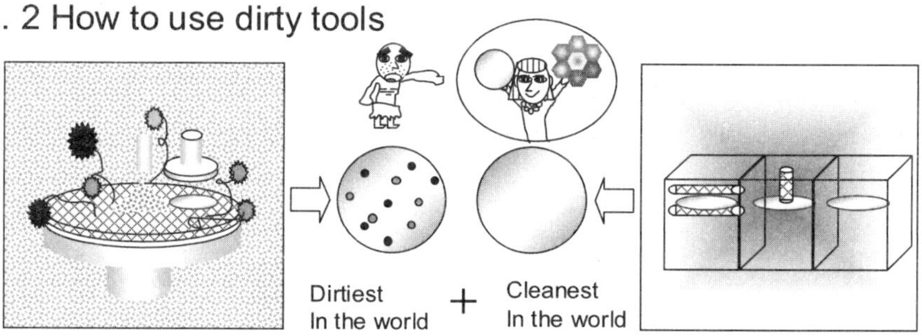

Dry-in and dry-out.
Solve by air-flow analysis

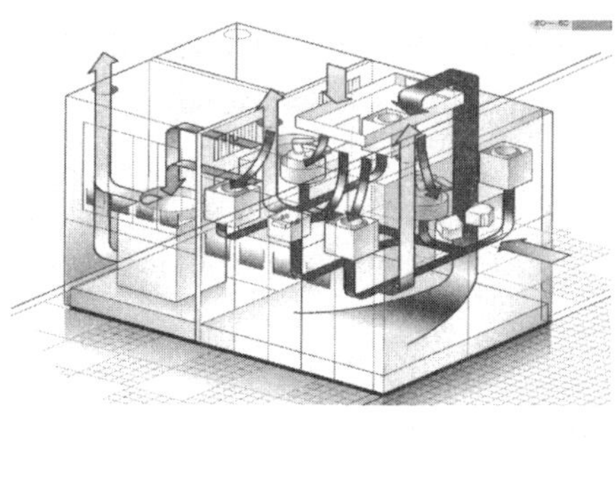

Competitive strategies for plating and polishing

Plating was introduced to the semiconductor industry in the same period as CMP.

It is convenient to compare plating and CMP based on competitive strategies, as proposed by Michael E. Porter [3).

Competitive strategies for CMP (Figure 1)

1. The CMP industry is *vulnerable* in terms of relationships with clients (device manufactures) since they are traditionally in a strong negotiating position. However, the market is an oligopoly, and the situation may gradually improve.
2. The CMP industry is *advantageous* in terms of relationships with suppliers (slurry manufacturers, etc.) since they are at present in a weak negotiating position. However, their negotiating position may be stronger in the future.
3. The industry is very *advantageous* with no alternative available for now.
4. There have been no newcomers so far because of the effect of patents.
5. The CMP industry is an oligopoly of the top two companies.

These considerations suggest that the situation of the CMP industry is quite advantageous.

Competitive strategies for plating (Figure 2)

1. The plating industry is *vulnerable* in terms of relationships with clients (device manufactures) since they are traditionally in a strong negotiating position. However, the market is an oligopoly, and the situation may gradually improve.
2. Suppliers (plating solution manufacturers, etc.) are currently in a strong negotiating position. However, as their negotiating position is weakened, the plating industry may become *advantageous*.
3. The industry is *vulnerable* with alternatives available (CVD, printing, etc.).
4. Newcomers may enter the industry since the coverage of patents is limited compared to the CMP industry.
5. There are many manufacturers in the plating industry.

These considerations suggest that the situation of the plating industry is less advantageous.

Plating and polishing were introduced to the semiconductor industry with similar backgrounds, but the difference between their states is apparent in terms of competitive strategies.

This fact may reflect that CMP is the only one technology, while plating is not.

Fig.1 Competitive strategy of CMP

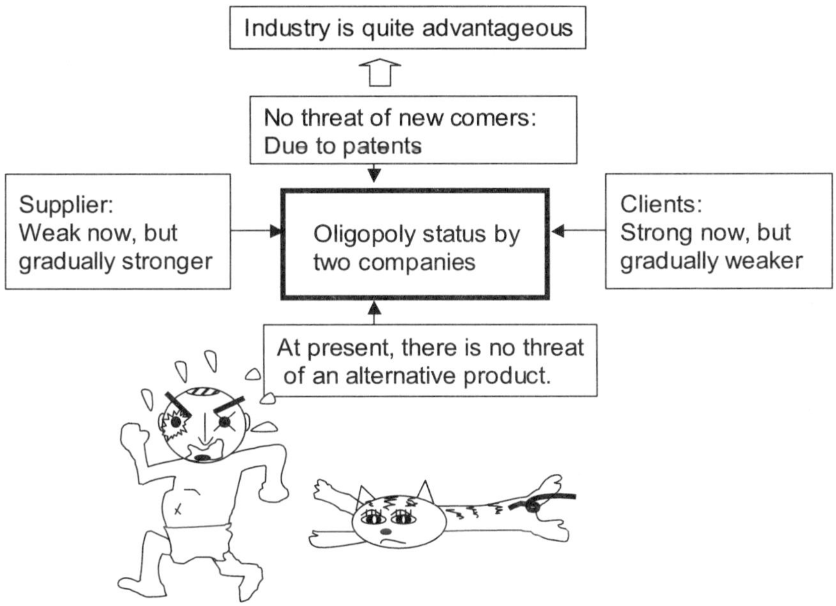

Fig. 2 Competitive strategy of plating

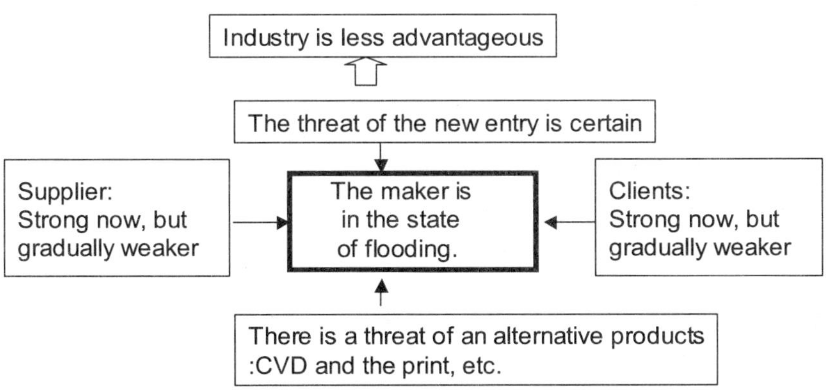

Better and must technologies

The second rule, "Develop must technologies, not better technologies," is as explained earlier.

Accordingly, we participated and succeeded in the development of dry pumps and CMP systems indispensable for semiconductor equipment. I believe that they have improved the performance of semiconductor devices, thus contributing to the company and society.

This rule is applicable to all fields, not limited to research fields.

Here are questions. Please ask and answer them yourself.

What are you currently **researching**?
What happens without the item you are researching?
Is it a must item, not a better item?

What are you **designing**?
What happens without the item you are designing?
Is it a must item, not a better item?

What are you **manufacturing**?
What happens without the item you are manufacturing?
Is it a must item, not a better item?

What are you **selling**?
What happens without the item you are selling?
Is it a must item, not a better item?

How are you doing **financial management**?
What happens without doing financial management?
Is it a must item, not a better item?

Do you feel that you are stuck at work (**routine work**)?
What happens without doing it?
Is it a must item, not a better item?

Honestly speaking, the second rule is the most difficult to fulfill.

R&D
It may be OK, even if there is no CMP. After all, is this 'Better'?
Design
It cannot be said that this is only one technology.
Production
If there is no screw, no machines would not work.
Finance
The company does not work if there is no money. (However, this :.)
All your business
Even in your life?
Memo
Boss, you cannot do anything if I don't help.

Just breathing...

"Think development as a cost that must be recovered" (Figure 1).

This rule advises that profitability (ROA, ROE, and ROI) be considered. Development must be performed with a management perspective. Take profitability into consideration. I used to be told that just breathing costs money.

Launch of development projects (Figure 2)

A development method widely used since the 1980s is presented below. The method below is useful in all ages. Companies tend to be concerned about the profitability of projects before launching them. As a result, they often make low-risk choices and miss huge business opportunities.

This problem can be solved by establishing corporate projects. The entire cost of each project is taken care of at the corporate level to facilitate challenges. This method emphasizes the following points:

- Seizing business opportunities;
- Managing the cost at the corporate level, and;
- Leaving all development issues to the project leader.

This method offers no way out for development engineers!!

Realistic planning (Figure 3)

If the cost is taken care of at the corporate level, can any kind of projects be launched? Of course, not. Project plans may be refused depending on their contents. It is important to realistically plan projects with some ambition.

For a company with a sales revenue of 100 billion yen, for example, the following three criteria are recommended for judging project plans.

- Possible market size of 10 billion yen (10 % of the company's revenue)
- Development of products that are *not easy to develop*
- Feasibility of the company's own technologies

Projects satisfying these criteria are green-lighted!

Fig.1 Recover R&D cost.

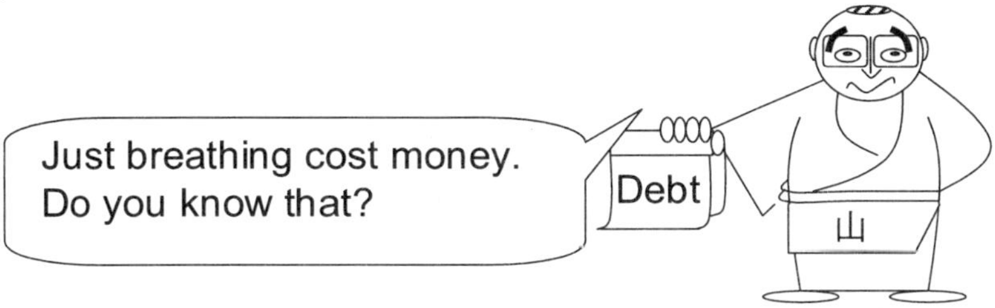

Fig. 2 Corporate project

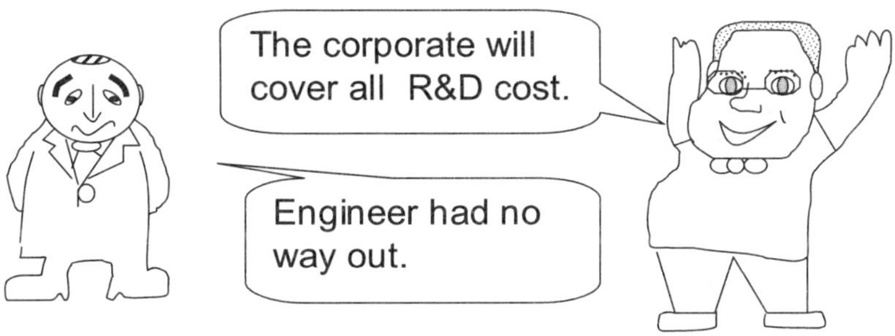

Fig. 3 Realistic planning for you

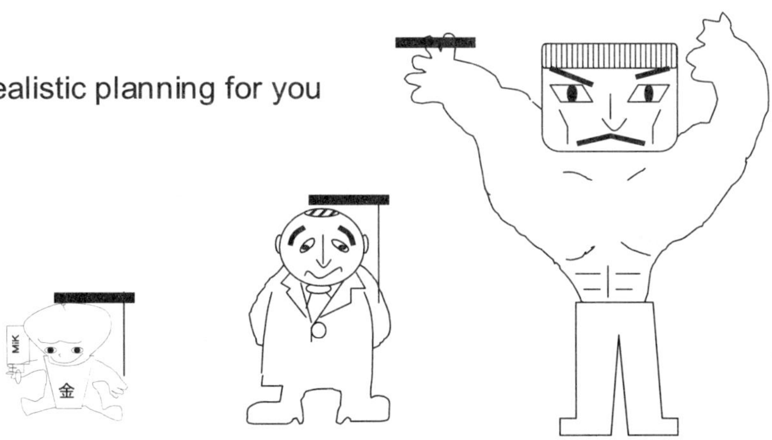

Why not breathe deeply?

Case study: Development of semiconductor equipment (Figure 1)

Here is an example from Ebara. In the early 1980s, Ebara launched some corporate projects. One of them was the development of semiconductor equipment.

The goal of the project was to *develop both markets and products.* In other words, we had neither markets nor products in the semiconductor field. We only had plans. Therefore, we established three criteria as mentioned earlier:

- Possible market size of 100 billion yen;
- Development of products that are *not easy to develop*, and;
- Feasibility of the company's own technologies.

The semiconductor project met all the criteria and was green-lighted!

Breathing deeply, we launched the project.

Food chain in the semiconductor industry (Figure 2)

The food chain in the semiconductor industry is simple. The electronics market is on the order of 100 trillion yen. Generally, semiconductor devices account for 20 % of components in electronics, and the market is on the order of 20 trillion yen. Device manufacturing equipment accounts for 10 % of investments by device manufacturers, and the market is on the order of 2 trillion yen. A ratio of 100:20:2 represents the semiconductor industry's food chain.

The food chain is dynamic. A 2 % change in the electronics market leads to a 20 % change in the device market and further to a 50 % change in the equipment market. This phenomenon is sometimes called the *bullwhip effect.*

Battle royal between three oligopolists in the semiconductor equipment market (Figure 3)

The competition in the semiconductor equipment market is fierce. The top two companies can survive, but the others are difficult to do so. At first, more than 20 companies competed in a battle royal; finally, only a few companies can survive. For example, the market of etching or CVD equipment has a size of about 300 billion yen. Since the equipment costs about 300 million yen per unit, about 1000 units are produced annually. If the top manufacturer has a market share of 50 %, it produces 500 units with sales of 150 billion yen.

You may feel timid if you are told that breathing costs money, but it is possible to develop a system and earn 150 billion yen. Why not breathe deeply?

Fig. 1 The corporate project starts.

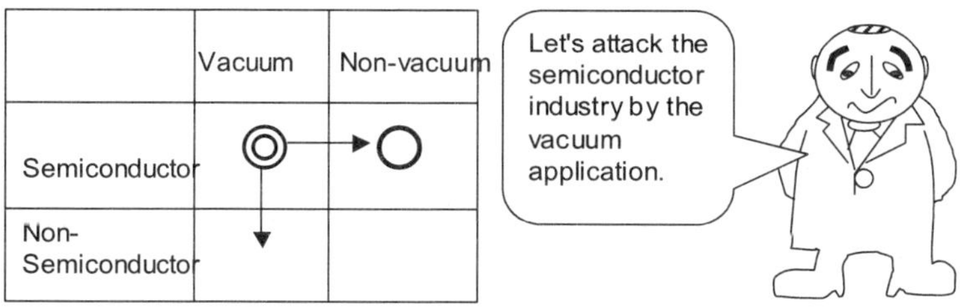

Fig. 2 Food chain of Semiconductor field

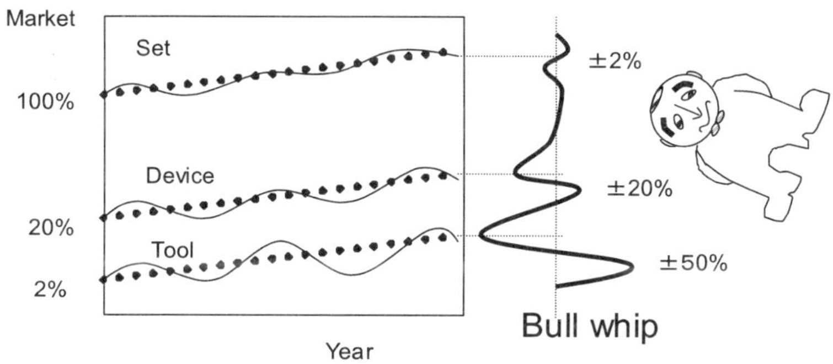

Fig.3 Semiconductor business is like 'Battle royal'

IT, BT, NT and PL, BS, CF

The third rule, "Think development as a cost," is as explained earlier.

Accordingly, I have conducted development with a management perspective by always considering the recovery of costs, although this attitude may be common in the world...

Do you know IT, BT, and NT? (Figure 1)

If you are an engineer, you can easily answer this question.

IT: Information technology

BT: Biotechnology

NT: Nanotechnology

Don't say *it*, *but*, and *not*!

Do you know PL, BS, and CF? (Figure 2)

Engineers may think that PL is an abbreviation for product liability. However, look at the question from a management perspective.

PL is an abbreviation for profit-and-loss statement, which is one of three basic financial statements and shows a company's business results based on the difference between revenues and expenses.

BS is an abbreviation for balance sheet, which is one of three basic financial statements and shows a company's financial conditions, including assets, debts, and capitals.

CF is an abbreviation for cash flow statement, which is one of three basic financial statements and describes a company's flow of cash.

Engineers must comprehend these basic financial statements.

Questions (Figure 3)

What are you currently **researching/developing**?

How much personnel and cost are required annually?

What are the timing and amount of sales?

Have you ever described these values in the three basic financial statements (or at least PL)?

These values represent the cost of *breathing*.

Fig. 1 IT/BT/NT

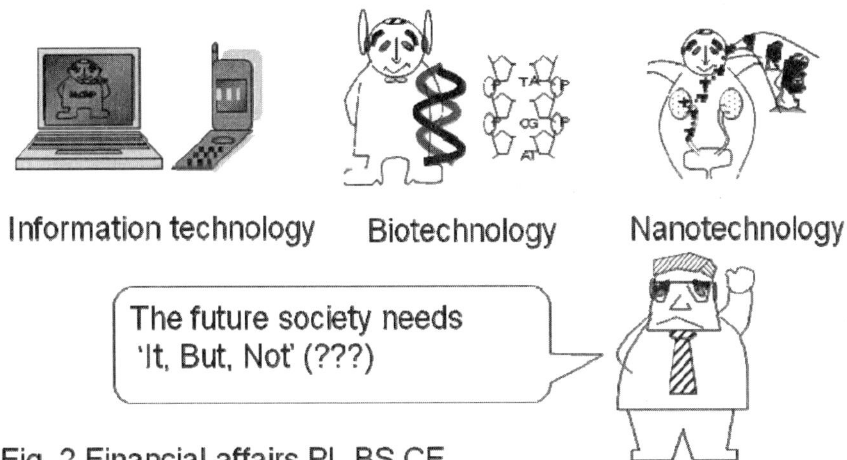

Fig. 2 Financial affairs PL.BS.CF

PL

Subject
Sales Operating expenses Operating profit Recurring profit Profit at this season

BS

property	Debt and capital
Current assets Fixed assets	Debt
	Capital
Assets	Debt & Capital

CF

Sales activity CF Investment activity CF Financial affairs activity CF

Fig. 3 Price of your R&D

Calendar	1	2	3	4	5
Investment					
Man	OO	OO	OOO	OOO	OO
Material	OO	OO	OOO	OOO	OO
Money	OO	OO	OOO	OOO	OO
Gain					
Sales			●	●●●	●●●●
Profit				●	●●●
Balance	–	–	–	–	0

Doing what you can do with some ambition

"Start with possessed technologies (use your expertise)."

This rule gives a warning on reckless diversification. My experience shows that Japanese companies are not good at M&A-based development. The companies should promote diversification and globalization using their own technologies. They have established clustering strategies for development and business though collaboration with other companies and organizations. This approach may be an ideal way for the Japanese version of Open Innovation[4)].

Starting with pump technology

Ebara has expertise in rotary machinery, especially in pumps. When entering the semiconductor industry, we decided to start with pump technology, particularly vacuum pumps. As mentioned before, a waste gas treatment system was necessary for developing a vacuum pump for semiconductor manufacturing. Thereafter, a clean vacuum system combining the dry vacuum pump and the dry waste gas treatment system was created.

Expertise in plating technology

Ebara also manufactures plating systems. A user knew this fact and gave us a chance to develop a gold plating system for device bumping (electrodes). In this way, a bump plating system was developed first, and then inquiries for various wet process systems were received.

Ultrapure water

Ebara is also known for ultrapure water technology. In the past, we had no share in the market of ultrapure water applications for semiconductor processing. Not just supplying ultrapure water as utilities, an on-site ultrapure water system supplying the required amount of pure water of required quality was developed.

Developing technologies one after the other

The development of the pure water system was followed by that of a cleaning system and then by that of a dry-in/dry-out CMP system combining the cleaning system and processes. R&D subjects increased, all of which originated from Ebara's basic technologies. An expansion of about 150 % was required for the development of semiconductor equipment; the hit ratio was about 20 %.

However, this level was sufficient to recover the investments.

Fig.1 History of R&D

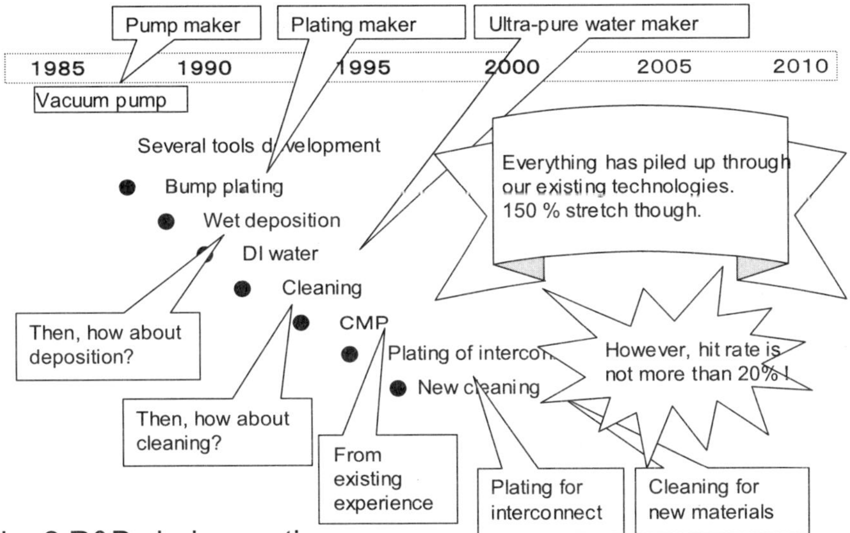

Fig. 2 R&D chain reaction

Line-up of semiconductor equipment

The fourth rule, "Start with possessed technologies," is as explained earlier.

Some promise to do what they cannot do. Their spirit of challenge is admirable but seems reckless in light of using their companies' money for development. It is advisable to start with your own accumulated technologies, while some ambition is required. Ebara has created many semiconductor systems and equipment. Some of them are presented below.

Chemical filter for lithography

Ebara has manufactured chemical filters using graft-polymerized membrane. Based on this technology, an ultra clean filter for semiconductor lithography was developed. It has earned a reputation as an innovative filter that can control the ppb order of amine.

Exhaust system for dry process equipment

As already explained, this clean vacuum system is based on the combination of a turbo molecular pump for ultrahigh vacuum, a dry pump for middle or low vacuum, and a waste gas treatment system.

Wet process systems

Wet process systems include CMP, plating, and cleaning. After a plating system was developed, a CMP system based on 3K polishing technology was developed. Then, polishing and cleaning were integrated to create a dry-in/dry-out CMP system (this is the greatest invention of the century, isn't?).

Utility system

Ebara is originally an ultrapure water manufacturer and can make ultrapure water systems for semiconductor manufacturing. To further meet users' needs, an on-site ultrapure water system supplying the required amount of ultrapure water of required quality was developed, instead of large-scale ultrapure equipment. Then, we further developed chemical recovery equipment.

The questions below are for students. What is your supervisor currently researching?

Do you think that the supervisor is well experienced?

Do you have some ambition now?

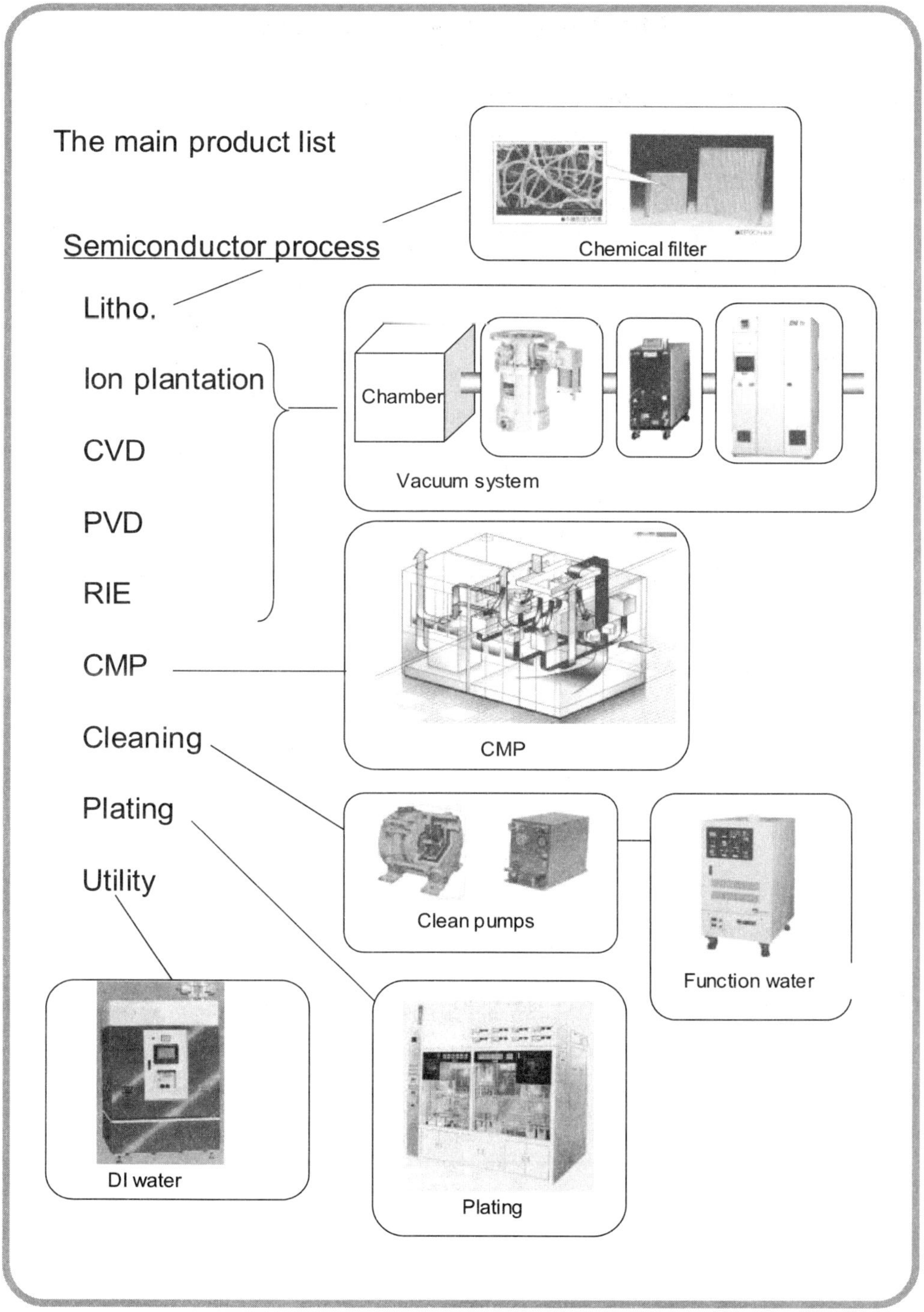
The main product list
Semiconductor process
Litho.
Ion plantation
CVD
PVD
RIE
CMP
Cleaning
Plating
Utility
Chemical filter
Chamber
Vacuum system
CMP
Clean pumps
Function water
DI water
Plating

Thinking until you feel sick

"Think and think! Think until you feel sick!"

This rule expresses a mental attitude that should be experienced at least once in achieving the goal of being a development-based company.

In fact, I have experiences of feeling sick when thinking.

When thinking of development for 24 hours a day, 7 days a week, and 365 days a year, the head may nearly explode. I experienced that my head heated up (about 0.5 °C?), and I became sick. When I go to the company by way of Fujisawa Station on the Tokaido Line, I have often felt sick (not drunk) and left the train at Totsuka Station before arriving at Fujisawa Station.

Nowadays, this extreme advice may be criticized as *bullying*. I no longer strongly advise anyone to do so. I wish that you remember this topic as an episode the author experienced.

The experience of thinking until feeling sick will last a lifetime!

When I spoke my experience to a researcher (VP class) at the IBM Yorktown Research Laboratory, the researcher said, "There is a slogan ‘Think Twice’ in IBM. IBM employees think only twice, but you think three times and further think until you feel sick. That's worth introducing."

When the *five rules* for development were presented during lectures at partner companies, universities, and high schools, I appreciated that some attendees felt empathy for my experience of feeling sick.

Have you felt the atmosphere of development at a company?
You do not have to follow the five rules.
I hope that you make your own five rules.

Go though your career with development-oriented thoughts anytime and anywhere.

Fig.1 My head drove recklessly by thinking too much.

Fig.2 IBM will think twice to get 'Out of box'.
I would think three times.

100:20:2 and 3:4:5:1

100:20:2

The semiconductor industry has the following food chain. Taking the electronics (chipset) market as 100%, the device market is about 20% (semiconductor devices account for about 20% of components in electronics), and the equipment market is about 2% (10% out of the 20%) (semiconductor equipment accounts for 10% of investments by device manufacturers). The electronics market rapidly grows with the emergence of new software applications; in turn, the innovation of semiconductor devices expands the range of software applications. Thus, the interaction between hardware and software drives the market growth. The demand for consumer electronics has no limit and may further increase in the future. Since higher-end products require a higher proportion of devices to be mounted, the needs for semiconductor devices may also continuously grow in the future.

3:4:5:1

The chipset, device, and equipment industries seem promising, but there is worrisome data. The average annual growth rate for 2008 and later is estimated to be over 3% for the world GDP, over 4% for the chipset industry, over 5% for the device industry, but only 1% for the equipment industry. Generally, an industry with a growth rate lower than the GDP is declining. Such an industry is not even in the maturity stage!

There is a Japanese comedian who has created a sensation by counting up from 1 and making a funny face when calling out multiples of 3. I have also created a new gag; that is, I call related figures and make a funny face when the figure is smaller than the preceding one, like 3, 4, 5, 1 (making a funny face).

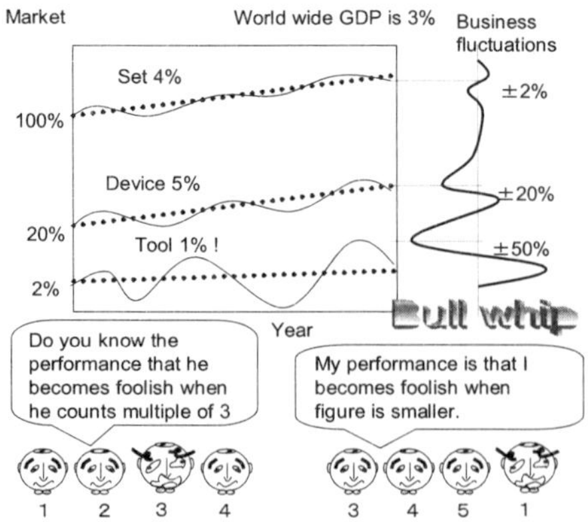

Chapter II

JOY (Defy Common Wisdom)

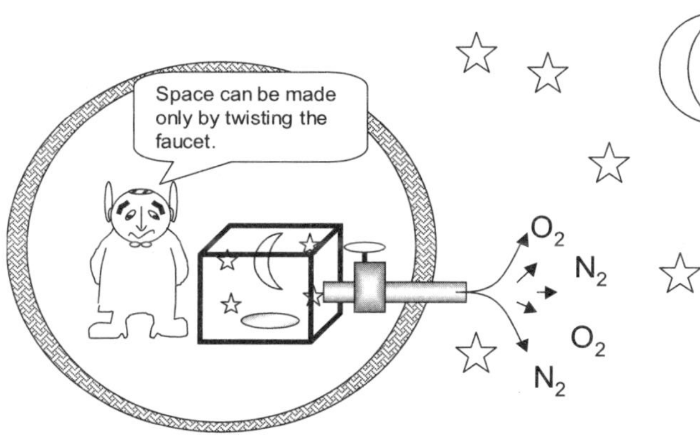

R&D in Space station

Eliminating unnecessary use - Mottainai!

Mottainai! (Figure 1)

With a view to achieve the ultimate goal of realizing a low-carbon society, saving is recently considered a virtue. However, our generation, baby boomers, has grown up in an era in which "consumption is a virtue." What is learned in the cradle is carried to the grave. The acquired habit is difficult to change.

I slept with the air conditioner on last night, reversing efforts to realize a low-carbon society. I regret that I did not turn it off when unnecessary!

I am depressed by finding many things wasted without any need.

The Hokkaido Toyako Summit promised to reduce CO_2 emissions by 50% by 2050, and I successfully cut waste at home by 50% in one day.

Vacuum system for semiconductor manufacturing (Figure 2)

I would like to talk about vacuum systems for semiconductor manufacturing facilities. Commonly, they operate with their pumps running continuously!

To defy the common wisdom, we started with a question, "Is continuous pump operation really required?"

As a result, we found that the time during which the reaction chamber for semiconductor manufacturing must be kept vacuumed was limited. Considering the possibility of stopping the pump when vacuuming was not necessary, we estimated that the pump running time could be reduced by 50 %.

It was a great finding. A project team was formed to start verification tests.

A system could be configured with no problem if the valve was opened/closed before/after vacuuming was required in the reaction chamber to ensure the vacuum level in the chamber remained unaffected. Furthermore, the supply of utilities (electricity, cooling water, nitrogen gas, etc.) to the pump was shut off during pump stoppage. Consequently, a total energy reduction of 75% was achieved compared to the original system configuration.

We never expected this result.

Unexpected advantages are often obtained in the course of development. In the field of R&D, they are called serendipity (or windfalls).

Fig.1 Stop if you do not need.

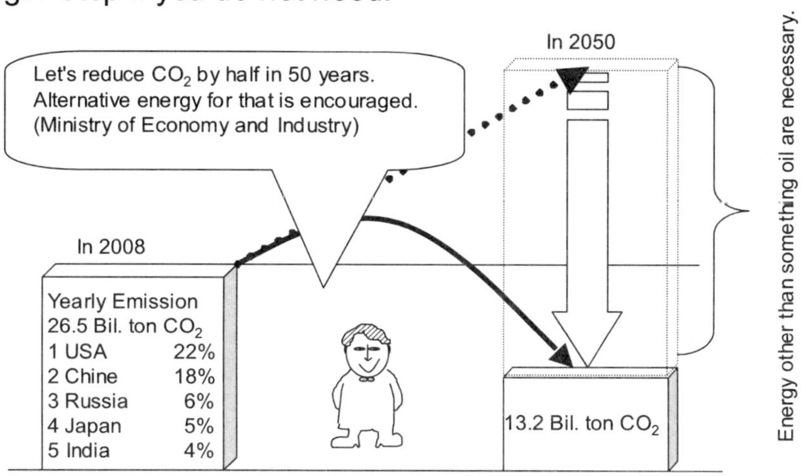

From Prime minister Fukuda speech on June 9,2008.

Fig. 2 We tried to stop it. So, what happened?

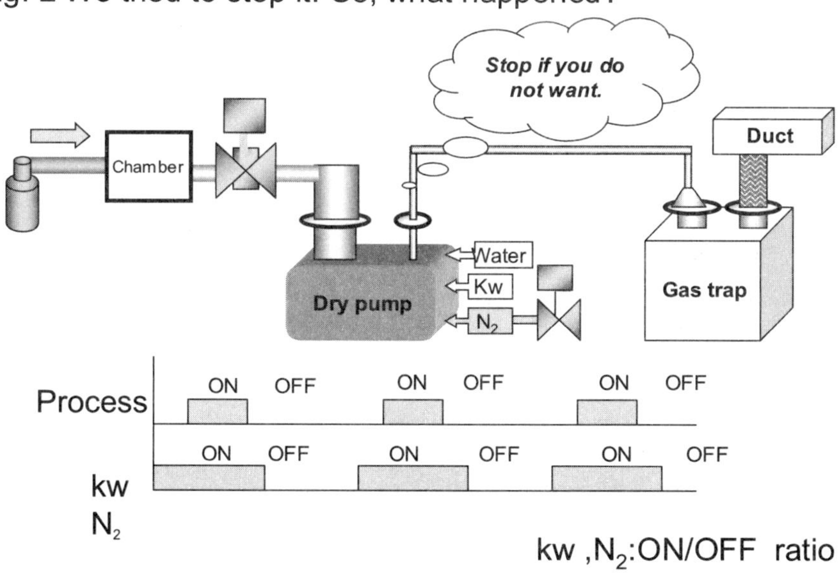

Removing the valves

Let's share! (Figure 1)

I share a small study room with Kumiko, my wife. This room is my favorite place in the small but comfortable house. The other day, Kumiko offered me an unbelievable proposal after watching a TV program with great interest.

The TV program was about how to optimize the use of limited space through space sharing.

The TV program presented many effective ideas, but Kumiko's proposal was different. Her proposal was to remove one of our desks, share the remaining desk, and use the space of the removed desk for her bookshelf. I would give way if I could use the half of the desk; however, according to the proposed non-aggression pact, I could use only 30% of the area on the desk.

This is not sharing but forcing!

Common wisdom of using valves (Figure 2)

Let's talk about vacuum systems for semiconductor manufacturing facilities again. See the right section of Figure 2.

In the conventional design, (1) a gate valve and (2) a pressure-regulating valve are provided separately between the turbo molecular pump and the vacuum chamber. This configuration requires a long distance between the ultrahigh vacuum pump and the chamber although it should be as short as possible. The configuration also significantly deteriorates the pump's vacuum performance.

Providing the valves (1) and (2) separately is common!

Why not integrate the two valves? In addition, they are installed inside the vacuum chamber (left section of Figure 2). A plate-like element in the vacuum chamber serves the functions of both the valves (1) and (2).

With a shorter distance between the turbo molecular pump and the vacuum chamber, the pump can achieve higher vacuum performance. In some cases, a smaller turbo molecular pump may be used.

Thus, space saving and improved vacuum performance can be attained at the same time.

Fig.1 Let's share.

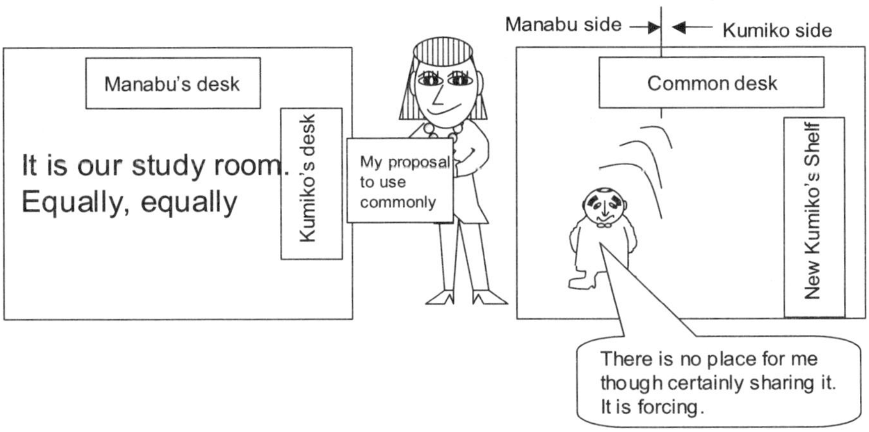

Fig.2 Integrate all features

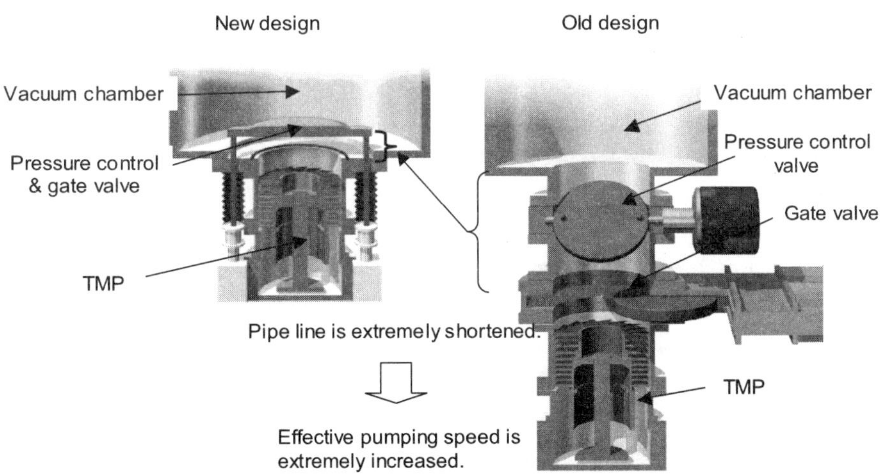

Only what is needed, as much as needed

Is it true that too big is better than too small? (Figure 1)

When I was frightened by the prediction of another Great Kanto Earthquake, I ran to buy emergency supplies. At the shop, an excessively smiling assistant came to me and said,

"You are lucky. We are holding a sale to celebrate the release of the movie *Nihon Chinbotsu* (Japan Sinks). You can get a bag of emergency supplies for a bargain."

"Uh… I don't need such a big one," I replied.

The shop assistant told me confidently, "Well, too big is better than too small."

"… OK, I will buy it."

When I brought it back home, I could find no space for storing the big bag. I wondered what Kumiko (my wife) would say when she came to find it…

We should buy only what is needed, as much as needed.

Oh, I know this expression. I used the same catch phrase for an on-site ultrapure water system, the development of which I participated in.

On-site ultrapure water system (Figure 2)

For clean room use, ultrapure water is generally supplied in large volumes by central systems. To meet more severe requirements on ultrapure water quality as the device generation advances, ultrapure water systems are substantially modified to improve the quality of all ultrapure water supplied in each plant.

It has been revealed that the proportion of processes requiring the highest quality of ultrapure water is not more than 20 %. Why not improve the quality of ultrapure water only for these processes? We started the development of an on-site ultrapure water system [5)] based on the concept of improving ultrapure water quality as much as needed for the relevant processes only.

Once the project started, it became clear that the required ultrapure water quality differed depending on the process. This finding led to the development of various types of functional water, such as ultrapure water with low dissolved oxygen, ozone-added ultrapure water, and ultra-purified ionized water.

Fig.1 Too big is better than too small

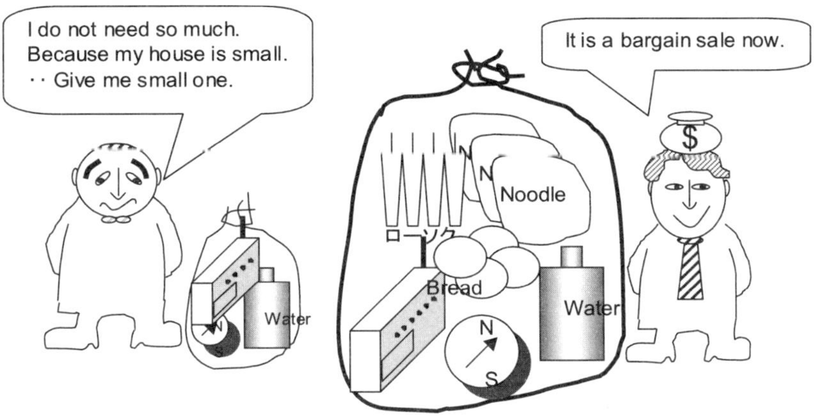

Fig.2 On-site ultra pure water system

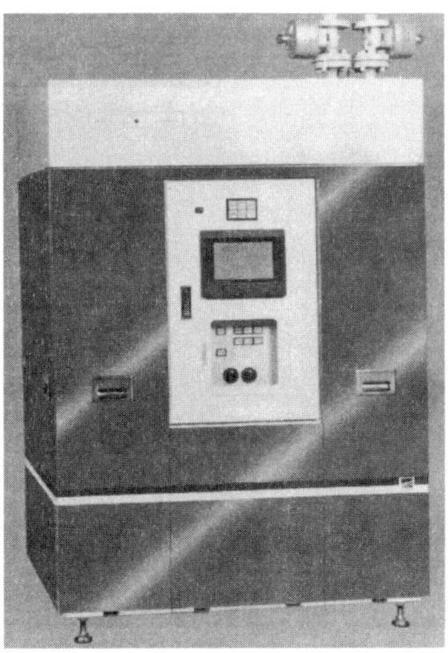

Improve performance only needed.
5 m^3/hr 10 m^3/hr

	Inlet	Outlet
Resistivity (MΩ·cm)	17.0<	18.1<
DO (μg/l)	< 100	< 5
TOC (μg/l)	< 30	< 5
Bacteria (counts /l)	< 50	< 1
Particle (counts /ml)	< 200	< 1
Silica (μg/l)	< 30	< 1

What is CMP?

Spock on a space station (Figure 1)

In 1986, I was occupied with working on ultrahigh vacuum.

At that time, I had the following dream.

I see someone testing a vacuum process on a space station.
He has pointed ears. Uh, he is Mr. Spock from Star Trek.
The inside of the vacuum chamber is vacuum, as is the case with outer space.
What type of vacuum pump is in use?
Oh, he just turns on the faucet, doesn't he?
Yes, all he needs is to turn on the faucet on the pipe connected to outer space, which is ultrahigh vacuum. He need not run a turbo molecular pump at 50,000 rpm like I did (I once did a destructive test for a turbo pump in an air-raid shelter).
He can easily obtain an ultrahigh vacuum, which is so difficult on earth…
A vacuum exists in outer space, and the atmosphere exists on earth.
I was inspired to develop atmospheric pressure processes on earth.

Developing non-vacuum processes, while vacuum is common (Figure 2)

I was given the opportunity to develop an eye-opening process when I was eager to achieve ultrahigh vacuum. The process scrubs semiconductor wafers with a cloth on a table while pouring liquid like creamed coffee! This was my first encounter with unbelievable CMP technology.

I knew that bare silicon wafers were finished by polishing. However, using a polishing system for wafer processing was a bolt out of the blue to me. Since the scaling and multilayer interconnection of semiconductor devices were required, as described earlier, the planarization of each layer was the top priority. Someone came up with an idea to develop a non-vacuum system for the reason that a vacuum was difficult to create (unfortunately, the idea was not mine, but was conceived by an overseas device manufacturer).

Nonetheless, it is Ebara that developed the dry-in/dry-out concept [6] for bringing the world's dirtiest polishing system into clean rooms for semiconductor device manufacturing, the world's cleanest environment. This technology has contributed to the creation of cutting-edge devices (we are proud of this fact).

Fig.1 The faucet is a vacuum pump in space.

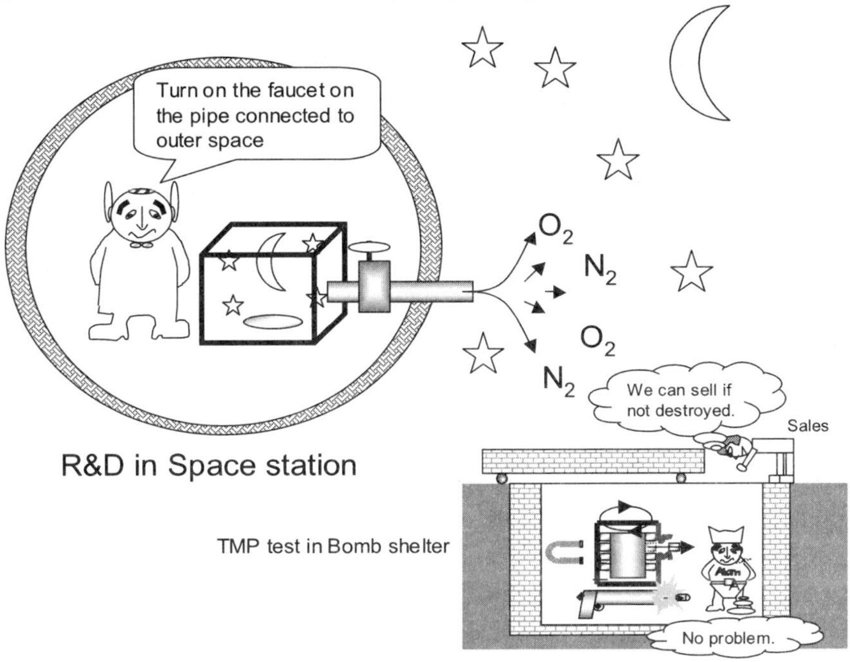

Fig.2 If vacuum is common wisdom, let's try non-vacuum.

Launch of the CMP development project

To let you feel the atmosphere of development, a CMP development project carried out at Ebara is introduced below.

Scaling and wafer size transitions for semiconductor devices (Figure 1)

The semiconductor industry has enjoyed prosperity with scaling and wafer size transitions since the transistor invention in 1947. Figure 1 schematically shows the evolution of semiconductors. As scaling proceeded, CMP was invented for semiconductor device planarization. For the 180 nm scaling generation and earlier, CMP was not adopted, and each layer was non-planar, as shown in the figure. For the 130 nm generation and later, CMP was gradually adopted; each layer was planar, and multilayer interconnection rapidly accelerated.

Disclosed development roadmap? (Figure 2)

Do you know that development specifications are open to the world in the semiconductor industry? Maybe because many semiconductor players support the concept of Open Innovation, device development specifications are disclosed in the International Technology Roadmap for Semiconductors (ITRS) [7].

How will devices evolve? What manufacturing processes are required for future devices? Semiconductor equipment manufacturers can set their own roadmaps for equipment development with reference to the ITRS and users' roadmaps.

For example, Figure 2 shows the progress of scaling and wafer size transitions with a time line. The ITRS also specifies other necessary parameters, relying on which engineers carry out development. However, it contains requirements, not solutions. Revolutionary technologies will be born from struggles to overcome technological limits.

The evolution of semiconductor devices triggered new material revolutions in device design and wet process revolutions in equipment design. CMP was devised through the same process.

In addition to equipment development for individual processes, *project management* covering multiple processes has been required. Project management [8] mentioned here is unlike that for NASA's big projects or public works projects. It aims to manage multiple teams with clear goals, such as performance improvement and cost reduction, in accordance with users' roadmaps.

Fig.1 Scaling and Wafer size transition

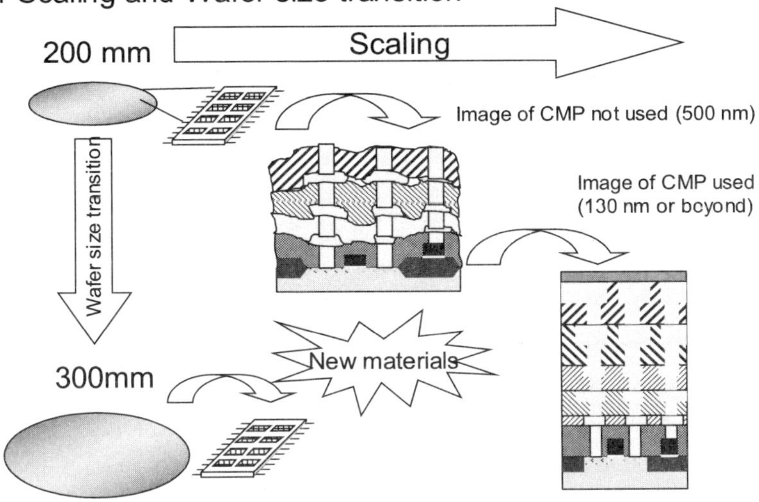

Fig.2 R&D schedule is disclosed

Year	2007	2010	2013	2016
Scaling (nm)	65	45	32	22
Wafer size (mm)	300	300	450	450

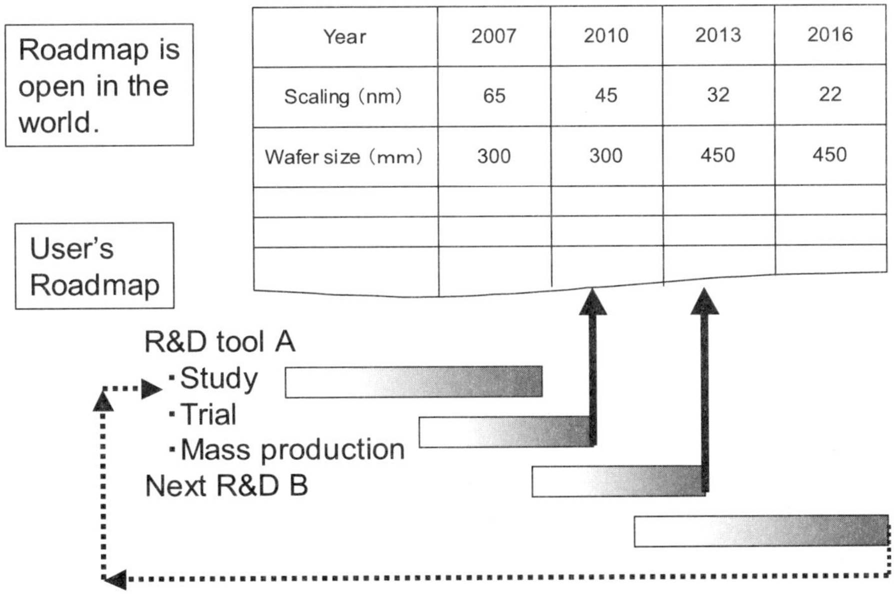

From dry to wet and then to CMP

Increasing use of wet process systems! (Figure 1)

The leader of the development team interpreted messages in the ITRS.

(1) Ultra-flat planarization: Planarity of 10 nm is required across an entire 300 mm wafer. This requirement is not easy to fulfill.

(2) Ultrahigh aspect ratio: As the line width becomes narrower, the line height increases. Consequently, the height-to-width aspect ratio increases.

(3) Ultrahigh-performance cleaning: Trench bottom cleaning becomes difficult, requiring higher cleaning performance.

(4) Adoption of new materials: New material options are required for semiconductor manufacturing.

(5) Wafer size transition: Wafer size transitions are required in all ages to improve productivity and cost efficiency.

Thus, manufacturing processes and related equipment must change substantially. To meet these requirements, a shift from conventional dry processes to wet processes has been made. Polishing and plating are examples of wet processes. The use of wet processes is expected to further expand in the future.

Outline of the damascene process (Figure 2)

This section explains the damascene process [9], revolutionary technology for semiconductor device manufacturing, in connection with plating and polishing.

Cu interconnects have replaced conventional Al interconnects because of lower resistivity. The damascene process is a technique in which copper film is formed by plating, and then the surface is planarized by polishing.

In Kyoto, there is a traditional craft called *zogan* (inlay), in which gold is inlaid by engraving a lacquered surface and pouring gold powder over the surface.

In the damascene process, oxide film etching is equivalent to engraving,
copper plating corresponds to the pouring of gold powder, and
trenches are inlaid with copper by polishing, instead of inlaying by hand.

Damascene technology for semiconductors is similar to the inlay technique.

Thanks to this technology, plating and polishing were introduced to semiconductor processing.

Fig.1 Wet tools and device roadmap

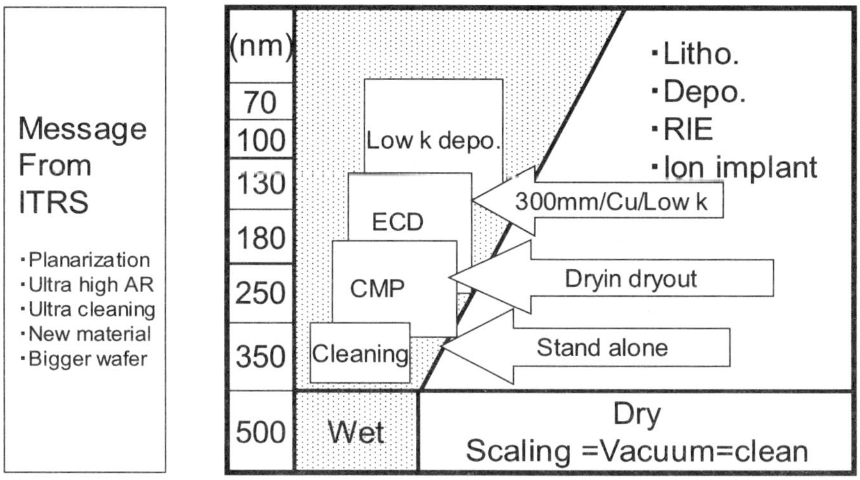

Fig. 2 Damascene process

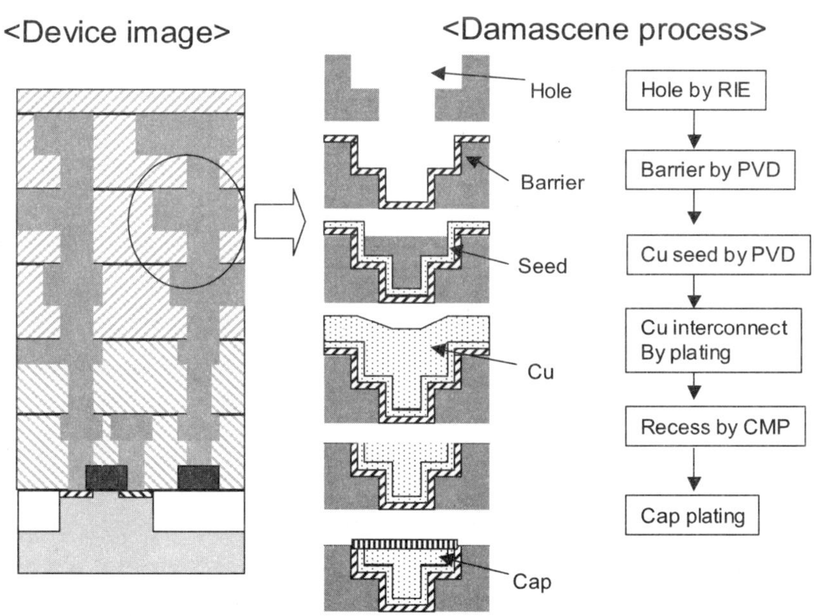

Overcoming challenges one after the other

Technological challenges presented by the ITRS (Figure 1)

The ITRS has presented technological challenges relating to the adoption of damascene technology as follows.

(1) The height-to-width aspect ratio increases with interconnect scaling.

(2) Interconnect barrier and seed layers become thinner.

(3) Polishing must provide higher planarization performance.

(4) An interconnect material with lower resistivity, i.e. Cu, must be adopted.

(5) Lower dielectric constant films, i.e. low-k materials and high-melting point metal caps, must be adopted.

(6) A wafer size transition is required.

Engineers around the world are competing to overcome these challenges. Besides plating and polishing, all technologies for successfully implementing the damascene process are under development everywhere in the world.

Progress of the project (Figure 2)

In the project, development concepts were set based on market and technical information in accordance with the development doctrine. Research institutes, the company's other divisions, universities, organizations, and concerned parties could be invited to the project. Project teams were formed according to respective concepts, and tests were done for each element. An in-house test version (α) was made based on the test results; a user evaluation version (β) was made based on the test results of the α version. Then, the final version (γ) was made. It was permitted to return to any project phase anytime.

Element technologies required for a damascene module include etching and sputtering (dry) and plating, polishing, and cleaning (wet). For the latter three, we aimed to enhance wet process equipment for semiconductor device manufacturing. Our target was to solve the technological challenges presented by the ITRS. The Wet Revolution means to overcome such challenges with wet processes. A common problem associated with wet processes is 3K: *kitanai* (dirty), *kitsui* (tough), and *keiken-izon* (experience-dependent). After defining how to enhance each element, development was done by each group. Finally, all elements were integrated. The key to project management is to "break down objectives into keywords or catchy words." Like NASA's word "we choose to go to the moon," objectives must be clarified to all team members by simple words.

Fig.1 Red brick shown by ITRS

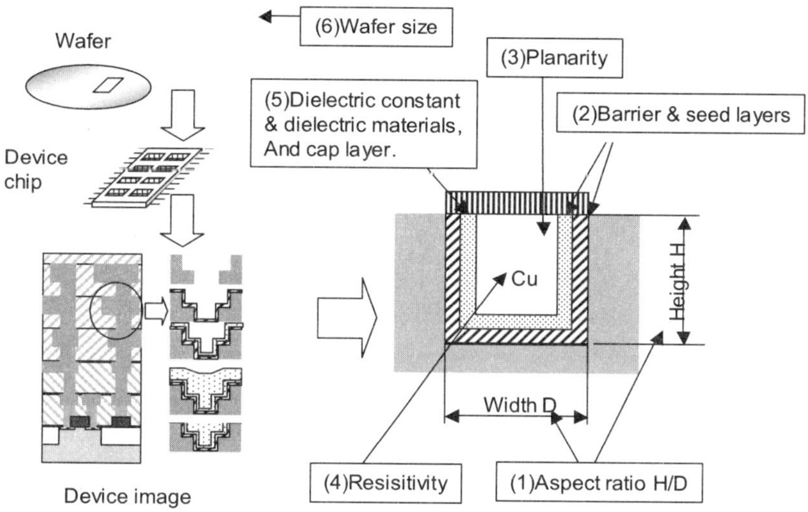

Fig.2 Project approach

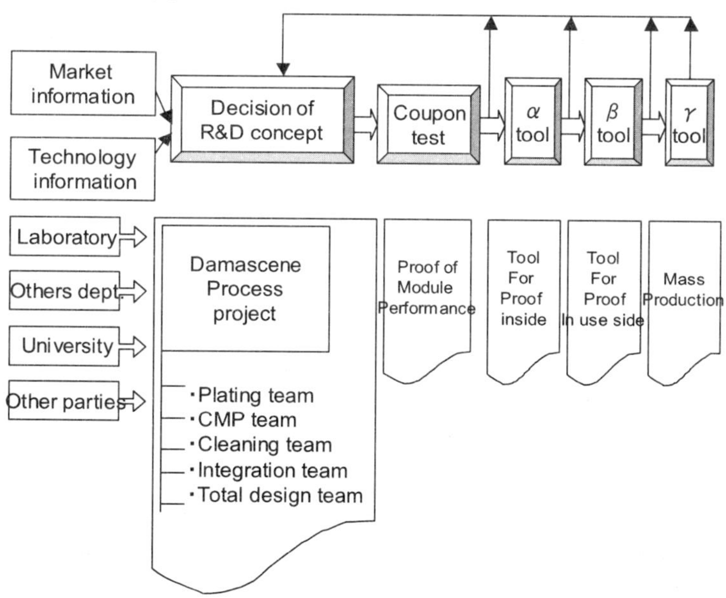

Project management

An example of project management [10] is introduced below.

Project policies

The figure shows the project's policies and challenges. The public acceptance of CMP, requirements for low-temperature and cost-efficient processes, and process changes with the adoption of new materials were identified as the project background.

The keywords for development were [process enhancement], [higher cleaning performance], and [scientific approach]. The project's targets were [lower resistance interconnects], [higher aspect ratio], [thinner barrier layer], [dielectric films with lower dielectric constants], [planarization], and [wafer size transition], as specified by the ITRS. Based on them, development policies and challenges were determined. We focused on process enhancement for plating and CMP. For plating, plating methods using dummy resistance and using a cap layer for Ag interconnects were chosen. For CMP, analyses for polishing rate distribution control and nanotopography were chosen.

Regarding higher cleaning performance, we focused on the enhancement of cleaning performance. We developed cavitation jet cleaning, dissolved oxygen removal, and watermark minimization with spin/vacuum drying. For details, see the references.

3K problems

Plating and polishing are typical wet processes. They have long been despised because of 3K problems, *kitanai* (dirty), *kitsui* (tough), and *keiken-izon* (experience-dependent). However, their performance was superior to that of conventional dry processes, steering engineers toward development with them.

Whenever I think of plating/polishing development, I am reminded of the proverb, "If a man keeps cherishing his old knowledge, so as continually to be acquiring new, he may be a teacher of others." The proverb is a sort of bible for development engineers. As described later, one of our project objectives was to reborn plating and polishing as new 3K processes: *kirei* (clean), *kantan* (easy), and *kagakuteki* (scientific). Our patented dry-in/dry-out wet processes were created with this concept.

Project strategy and several problems

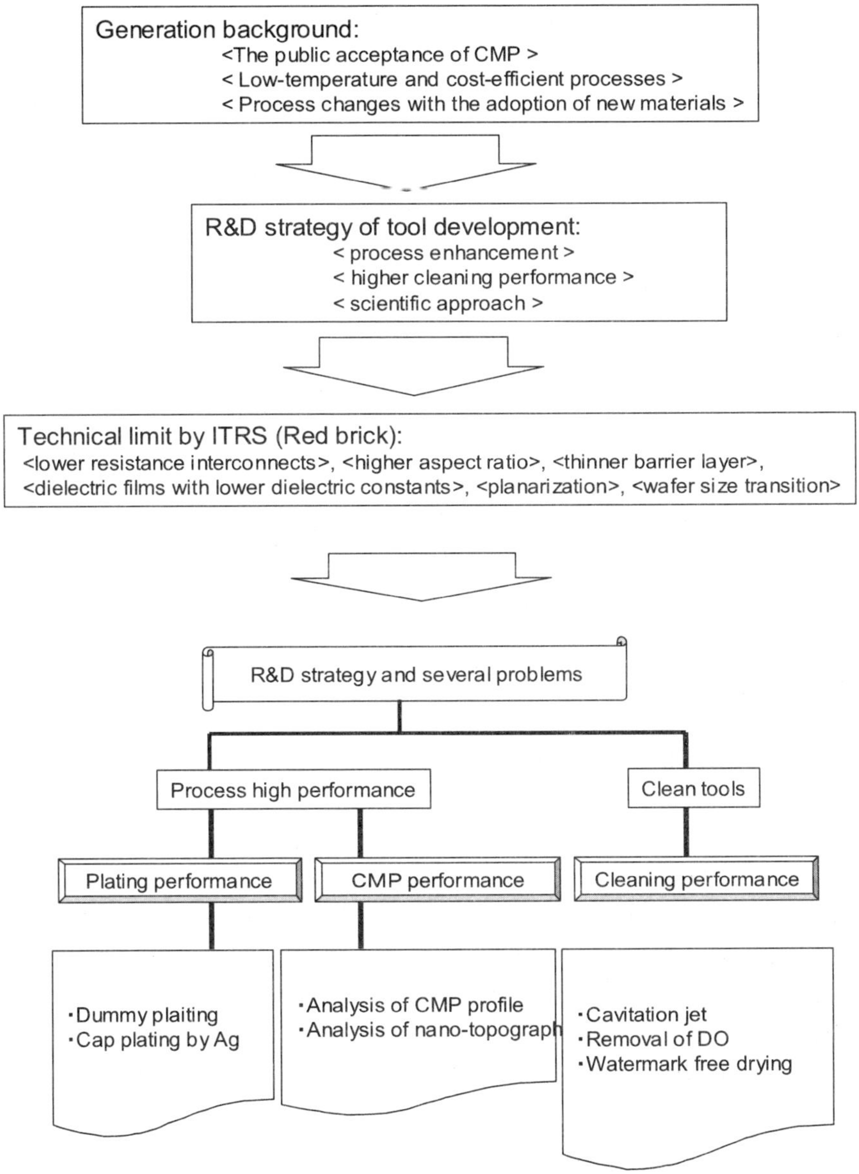

From the plating team

With a catchy slogan (Figure 1)

The late U.S. President John F. Kennedy reaffirmed, “we choose to go to the moon,” in promoting the nation’s space development. His slogan was powerful, easy to understand, and hopeful. A catchy slogan can be so effective.

We call polishers CMP. Am I the only person to feel that CMP sounds more sophisticated than polishers? Similarly, we have decided to call plating ECD (Electrical Chemical Deposition). It sounds more high tech, doesn’t it? This phrasing is an example of word magic and demonstrates the effect of catchy slogans. It may help change engineers’ way of thinking.

Conventional plating system (Figure 2)

In terms of plating, the project aimed to change the interconnect material from Al, conventional material, to Cu and to fill high-aspect trenches while using a thin seed layer. In a conventional plating system, a large amount of plating solution in a large circulation bath is circulated by a pump, as shown in Figure 2. This configuration poses many problems, including large system size, use of a large amount of plating solution, and bubble formation due to circulation.

We found solutions to these problems.

Bath of only 60 cc? (Figure 3)

This system has been developed to solve the problems above.

Instead of circulating plating solution, the system pours only 60 cc of plating solution over wafers. It has embodied the process of dummy plating (explained later as the God with Metabolic Syndrome). Dummy plating has more advantages. The scaling of semiconductor devices leads to a thinner seed layer, resulting in an increase in seed resistance during electroplating. Dummy plating offers successful plating deposition on a thin seed layer by providing extremely high dummy resistance in the electrical system for plating.

Why?

Please wait a while. The God with Metabolic Syndrome will appear soon.

Fig.1 Catch slogan

Fig.2 Conventional plating systems

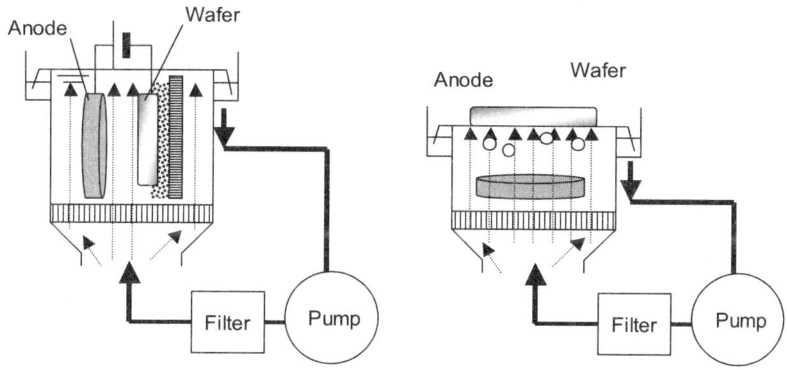

Fig.3 New dummy plating

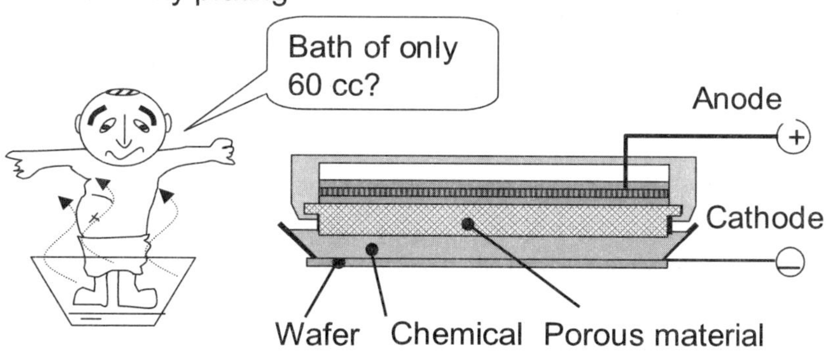

From the polishing team

Not polisher, but planarization (Figure 1)

Polishing has evolved in the form of CMP. Rephrasing polishers as CMP changed engineers' way of thinking. I tried further rephrasing. The "P" in CMP originally stands for polishers, but users do not demand polishers but ultra-flat planarization. The "P" need not stand for polishers any more, and CMP may be renamed. In fact, CMP was renamed from Chemical Mechanical Polisher to Chemical Mechanical Planarization in 1998. The new name has been adopted since MRS Meeting 2001.

General principle of planarization technologies (Figure 2)

What sorts of planarization methods are available other than polishing?

Before the emergence of CMP, some planarization processes, such as deposition by dielectric planarization and metal reflow, were used. However, they were gone after the introduction of CMP. As the next generation planarization systems after CMP, Electro-Chemical Polisher (ECP), Electro-Chemical Mechanical Polisher (ECMP), etching, and their combination are target technologies.

ECP can be regarded as the reverse process of plating. Generally, ECP has been proven effective for deburring. It has not produced the expected good results in planarization.

ECMP is based on the combination of ECP and polishing. Polishing uses an oxidizer for copper oxidation, while ECMP electrically performs oxidation. In both cases, oxidized copper is complexated by a complexing agent, and then mechanical removal is performed using slurry. The difference between the two methods exists in copper oxidation. Etching is performed to chemically remove copper.

Basically, polishing and ECMP have the same planarization ability. Since ECP and etching, in principle, have no planarization ability, other measures for planarization, such as addition of chemical agents, is necessary. These principles were summarized and published as a paper entitled *General Principle of Planarization Governing CMP, ECP, ECMP & CE* [10]. This paper has not focused on theoretical development, but has gained a reputation that it is useful for understanding the features of various planarization technologies.

Fig.1 Not polisher, but planarization

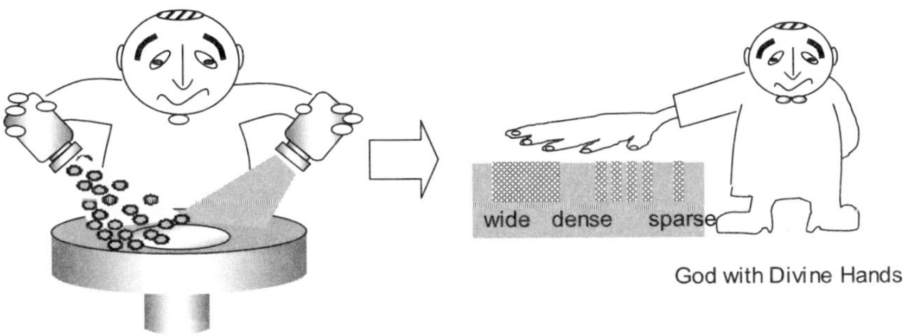

Fig.2 It is not necessary polisher

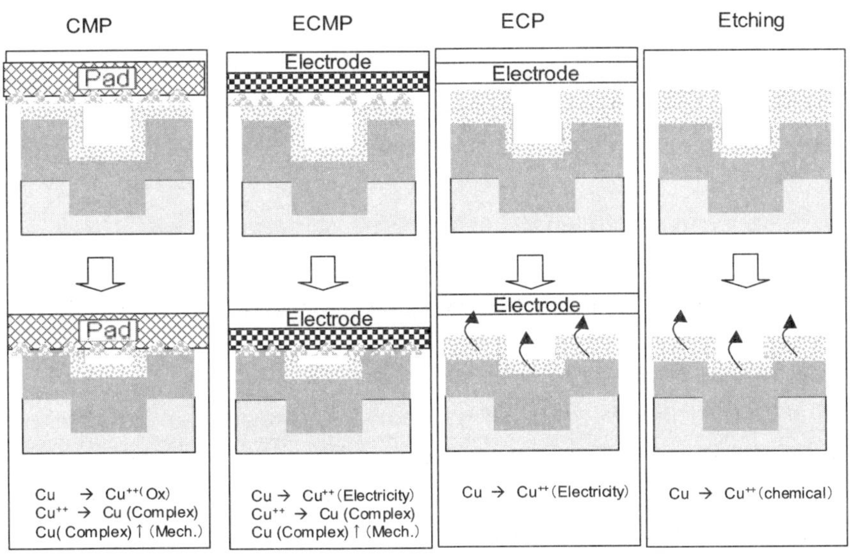

From the cleaning team

Development of dry-in/dry-out technology (Figure 1)

In the early days of CMP's history, a wrong belief that "wafers must be kept wet since dried slurry is stuck on them and cannot be cleaned" was widespread. In fact, the polishing and cleaning units were standalone and independent from each other, as shown in Figure 1. Kept immersed in water to avoid drying, polished wafers were transported to the cleaning unit by hand! It is no wonder that few people wanted to use CMP. How could we make CMP as user-friendly as dry process systems? It was desired to put wafers in dry state (dry-in), clean/dry them after polishing, and take them out in dry state (dry-out). This is called the dry-in/dry-out concept [11)].

However, people tend to have wrong beliefs.

Polishing was the world's dirtiest process, while cleaning was the world's cleanest process.

No one imagined that these two processes could be integrated.

Airflow analysis (Figure 2)

Is that true?

To answer the question, the project team did repeated tests and analyses.

First, airflow control was considered.

A promising approach was to control airflow from the cleanest load/unload subsystem through the cleaning subsystem to the dirtiest polishing subsystem. The dry-in/dry-out system was created through the repeated cycle of analysis with the finite element method (FEM), trial fabrication, and verification.

Without the dry-in/dry-out concept, CMP would not be as user-friendly as dry process systems and not be as widely used as it is today. Technology has won.

Performance as a cleaning system (Figure 3)

Cleaning was originally intended to remove slurry, but the demand for higher cleaning performance was growing. Semiconductor device cleaning is difficult; contamination must be avoided when contaminants exist on the same surface.

The project had many twists and turns; e.g. the progress speed varied between the plating, polishing, and cleaning teams, and some defects occurred due to the other teams' causes. The outcomes of the project showed the possibility to overcome *technological limits* in ECD, CMP, cleaning, and drying presented by the ITRS.

Fig.1 Development of dry-in/dry-out technology

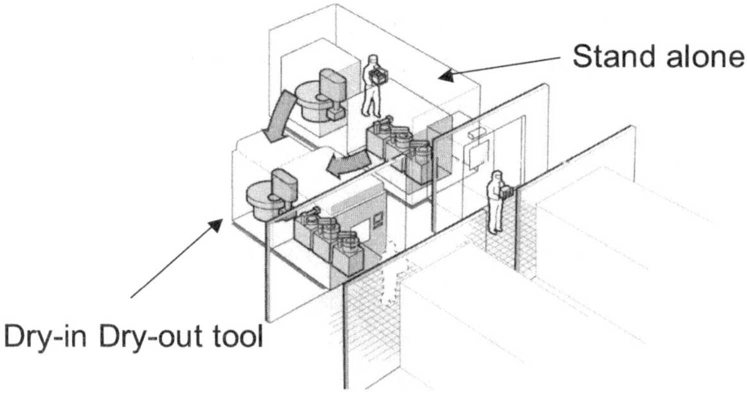

Fig.2 Air flow analysis is useful.

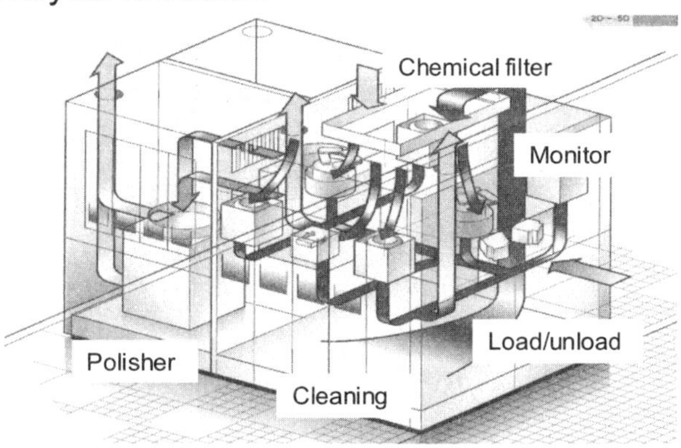

Fig.3 How to clean?

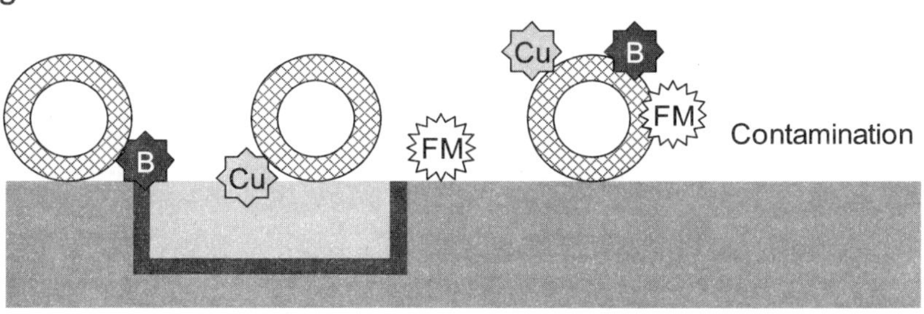

JOY

Joy! Joy!
We chant the word while conducting development activities.
Manabu: Let's think what common wisdom is.
High school student: Offering a seat to an elderly person on a train.
Manabu: Then, defy the common wisdom.
Student: Occupying the elderly person's seat?
Manabu: …
This is an actual conversation I had at High Tech University.
What was wrong?
"Offering a seat to an elderly person on a train is not a *common wisdom* but a *moral issue*.
Smoking cannabis or breaking traffic rules violates *laws*.
I am not talking about legal violations or moral issues. JOY stands for *Joshiki O Yaburo* and is intended to evoke joy by defying common wisdom that is practiced unconsciously as if it were water or air.
High school student: Getting up at 7:00 am is my habit.
Manabu: Then, defy the common wisdom.
Student: I will get up at 6:00 am.
What will happen to this high school student in the next morning?
I wish him a joyful discovery.

Chapter III

How to Survive Using R&D Concepts

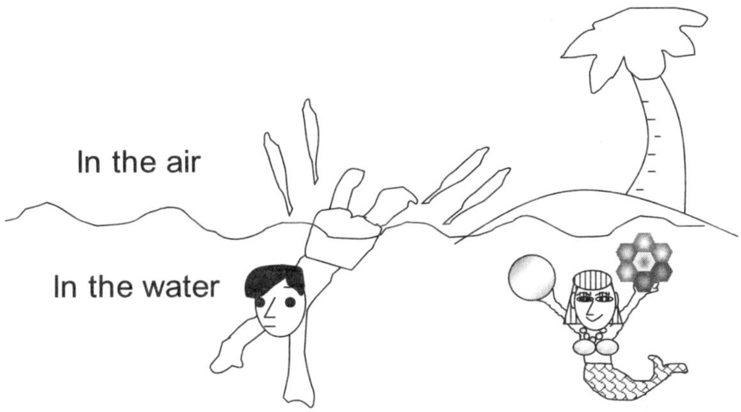

To be the world's No. 1

Which company? - A question to those making career decisions (Figure 1)

Here are questions to those making career decisions. If you are already working, answer them by recalling the time when you made the decision.
First question: Do you choose a large company, as the proverb says, "Look for a big tree when you seek shelter," or a small to mid-size company, as another proverb says, "Better be the head of a dog than the tail of a lion" (Figure 1)?

This choice appears to be largely influenced by the economic climate. When the economy is bad, a big tree is popular; when the economy is good, people tend to choose the head of a dog. Regardless of which type of company you choose, all companies have good and bad times. It would be no joke if your company were on its beam-ends in 30 years after you start working for it. Since no one can predict 30 years ahead, you should do what you like the most.

Everything has a limited life (Figure 2)

Industries have a limited life. The key industry shifted from steel through automobile to semiconductor; it may further shift to biotechnology, environmental business, or even agriculture. Every company has ups and downs according to the life of each industry. You may face a difficult time at least once in 30 years. If this cycle shortens, there will be a tragedy.

In conclusion, you may choose either a large or small to mid-size company. Now, let me make you feel like the world's No. 1.

How to become the world's No. 1 (Figure 3)

Another question: Regardless of the field, is there anyone who can say "I am No. 1 in the world" or "There is no rival in my field"?

Some may say, "I am the world's No. 1 rice eater." That is not too bad, but let's set a goal to become the world's No. 1 in your career life.

Similar to the theory that a company's life cycle is 30 years, the life cycle of an employee is 30 years. You want to be the world's No. 1 in 30 years, right? I have at least five fellow engineers who are the world's No. 1 in their fields.

Do you think that you will have a chance to be the world's No. 1 while working for a big tree or as the head of a dog for 30 years? Consider what you can do in 30 years. You will probably have at least three chances.

This book will help you to be the world's No. 1.

For your information, I feel confident that I can run faster than a cat.

Fig.1 Big tree or the head of dog?

Fig.2 30 years in several life times

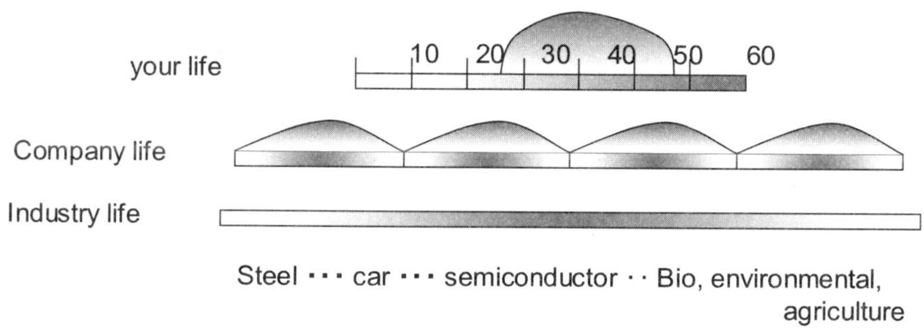

Fig.3 No.1 in the world

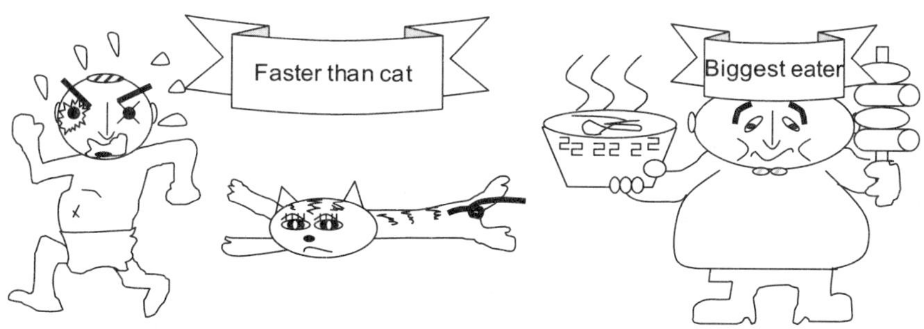

What is a development-based company?

What do you imagine about a development-based company? (Figure 1)

What do you imagine about a development-based company?

A company that develops the world's best PC?

A company that develops the world's smallest mobile phone?

Or, your company?

I think that every company should be development-based to survive this age. All companies, including barbecue sauce manufacturers, fast food companies, and mom-and-pop candy stores, must always think Something New for survival.

What is development? - Something New (Figure 2)

The first rule described in Chapter I tells us that development involves generating Something New ideas and taking actions to implement them. A company developing the world's best PC or a mom-and-pop candy store must always think new ideas and take actions to implement them.

Take chances and never be afraid of failure.

Of course, never promise what is impossible. Start with your own expertise, as stated by the fourth rule in Chapter I.

Review your situation according to the five rules again.

- Is your company challenging new possibilities?
- Is it a better idea or a must idea?
- Do you think that you can recover the cost of development?
- Have you started with your own expertise? It isn't a reckless plan, is it?
- Do you always think? Think and think…

If these checks are passed, your company is OK.

Example of model companies (Figure 3)

Here is an example from a company well known for its search engine. The company is located in the neighborhood of a hydraulic power plant, aiming to become a low-carbon company as a measure against global warming. Hydraulic power generation emits little carbon dioxide. There is little transmission loss between the company and the power plant. The company uses low-power servers. In addition, the company building has solar cells on its roof. Nothing could be better. This is also an idea to create Something New.

Fig.1 How about your company?

Fig.2 Follow five rules

1. Pursue Something New.
2. Develop *must* technologies, not better technologies.
3. Think development as a cost that must be recovered.
4. Start with possessed technologies (use your expertise).
5. Think and think! Think until you feel sick!

Fig.3 Do you know this company?

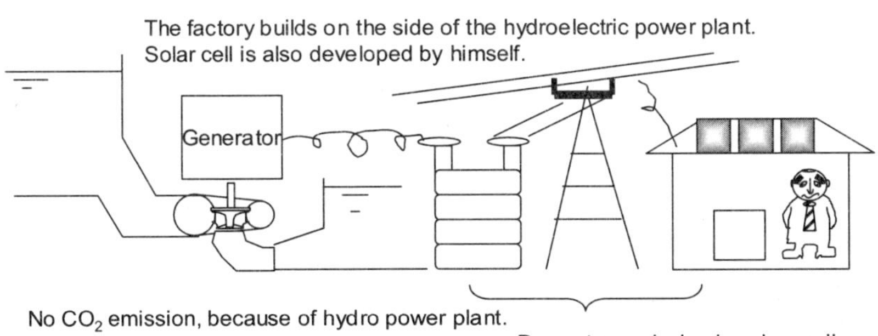

TKO - technical knockout?

Roles of TKO (technical knowledge officer) (Figure 1)

You want to be the world's No. 1.
Your company aims to become (or is already) a development-based company.
If these things are true, your company has pilots for development.
They are called TKOs (technical knowledge officers).
TKO does not mean technical knockout (though TKOs may get knocked out).
They may be called differently but are leaders responsible for development.
Development leaders must observe the three development rules below:

(1) Performance,
(2) Punctuality, and
(3) Price.

Poor performance is out of the question.
Development is wasted if it is not conducted timely.
Nobody buys a product if its price is extraordinarily high.
TKOs are responsible for all of these aspects.

Case study (Figure 2)

An example from Ebara is presented below. The company started technical and market developments at the same time. As a rotating machinery manufacturer, it had no expertise in vacuum technology, but was confident of developing vacuum pumps. Thus, Ebara kicked off the development of vacuum pumps for the semiconductor industry.

Later, Ebara had to choose either to target industries other than the semiconductor industry with vacuum technology or to develop technologies other than vacuum technology for the semiconductor industry. It chose the latter since requests for equipment development already rushed in from the semiconductor industry.

Making such decisions is another role of TKOs.

Conducting work on an organizational basis

I have written about the roles of TKOs, but actual development work is organizationally conducted by subordinates under TKOs. What happens if TKOs carry out whatever they like? TKOs should make an organization that has a function to check their own errors.

Fig.1 Role of TKO

TKO

(1) Performance
(2) Punctuality
(3) Price

Fig.2 Start from your own technology

	Vacuum	Non-vacuum
Semiconductor	◎ →	○
Non-semiconductor	↓	

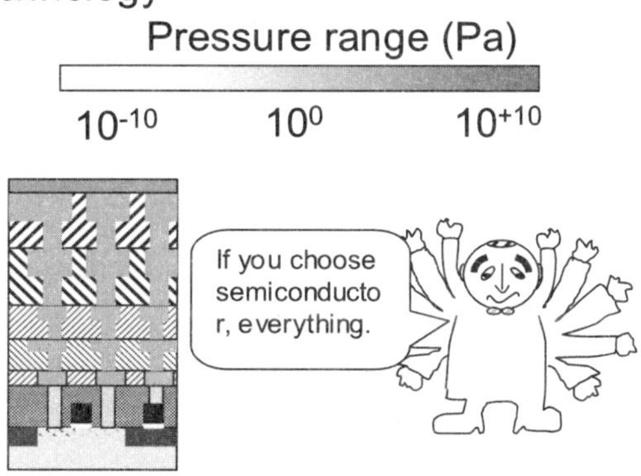

Fig.3 Fights in the organization!

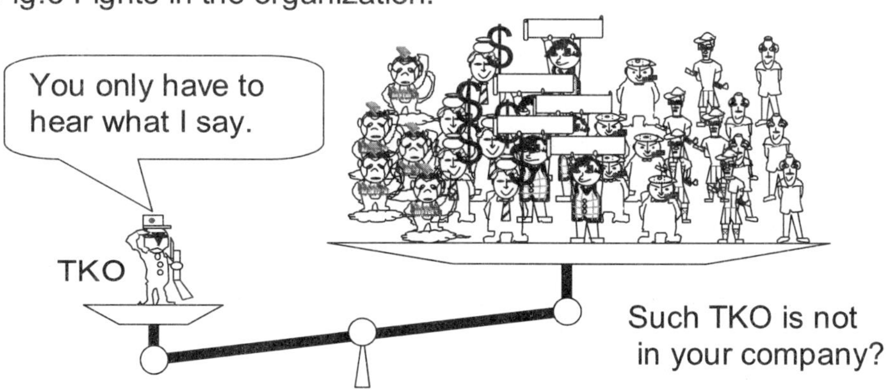

The case of Kumiko, secretary

I have told that no company or workplace could survive without R&D mind. To help you enrich your company life, some case studies are given below.

Close observation (Figure 1)

A secretary moaned:
"Our president is quick at work but a disorganized person. He keeps asking me, 'Where is this and that?' I spend most of the day looking for documents…"

Solving the problem after identifying it

The secretary tried to solve the problem.
When she organized her boss's papers the other day, she found that keywords were provided for all papers. She learned that they were important for later searching.
This is it!
An idea flashed across her mind.
She decided to ask her boss, "What is this document about? Please give me five keywords," whenever asked to organize documents.
It requires extra work, but a considerable number of documents have been organized since then.

According to the five rules

It's time for her to verify what she did according to the five rules.
- Something New: Keywords are commonly provided for papers, but not yet adopted for organizing documents.
- Is it a must idea, not a better idea?: It is still a better idea.
- Think development as a cost that must be recovered: Yes, my boss and I can avoid wasting valuable time.
- Start with possessed technologies (your expertise): I did not pursue secretarial studies, and I would like to receive professional training from now on. I regret that I could not do my work efficiently.
- Think and think! Think until you feel sick!: Not for me!

Fig. 1 Let's start from watching and thinking carefully

Fig. 2 Follow five rules

1. Pursue Something New.

Five key words are memorized.

2. Develop *must* technologies, not better technologies.

I cannot reach so high level, though.

3. Think development as a cost that must be recovered.

To save (Loss of my boss + my loss)

4. Start with possessed technologies (use your expertise).

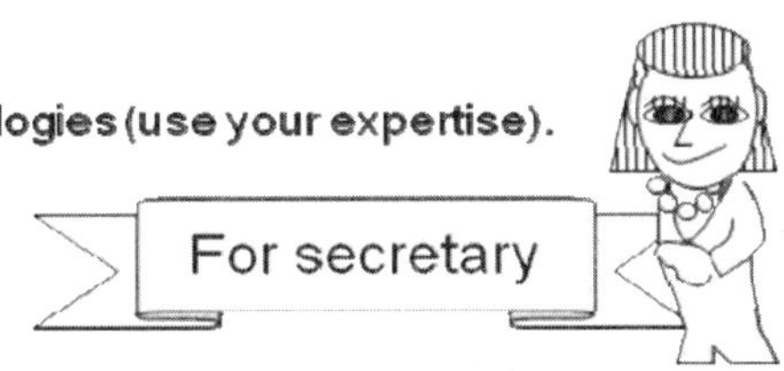

5. Think and think! Think until you feel sick!

The case of Mr. Yamada, newcomer to the personnel section

Yamada, newcomer to the personnel section, was dwelling on what he was told in Mr. Tsujimura's training program (Figure 1).

"Mr. Tsujimura told that all workplaces should be development-based. Is there anything I can do?"

Looking around with new eyes (Figure 2)

If you work in the same workplace for 10 years, you fall into a rut. Common practice for you may really be unreasonable. It is difficult to recognize that something you regard as common practice is unreasonable, like trying to feel air.

Newcomers, who may get shocked as if they jumped from the atmosphere into water, can see things differently. However, as they keep swimming, they will forget that they are in water. Yamada should take advantage of his freshness.

Internship program (Figure 3)

Every year, Ebara offers an internship program that spends much manpower and money. Yamada found that students participating in the program had no common interest with the receiving division.

Yamada said, "Boss, the students from the XX University seem to have no common interest with the YY division. In fact, the students are saying so."

His boss replied, "What? We did the same last year! Stop complaining and just do it!"

There is no boss like him, but his statement is enough to let Yamada down.

The department's manager took up Yamada's opinion and said, "Yamada, you make a good point. Internship programs are to match the interest of students and companies. You are right. We'll review the selection of students."

After the internship program

The three-month internship program was over.

Here is a letter from one of the participants.

"I have found that my research matches your company's business. I will study harder so that I can receive further guidance in your company next year."

Here is a comment by the department's manager.

"Our intern, XX, worked really hard. I hope he will join us next year."

Yamada said, "I could not follow the five rules exactly as taught by Mr. Tsujimura, but now I understand the necessity of generating new ideas with new eyes."

Fig.1 Something new even in Personnel section

Fig.2 If you look differently

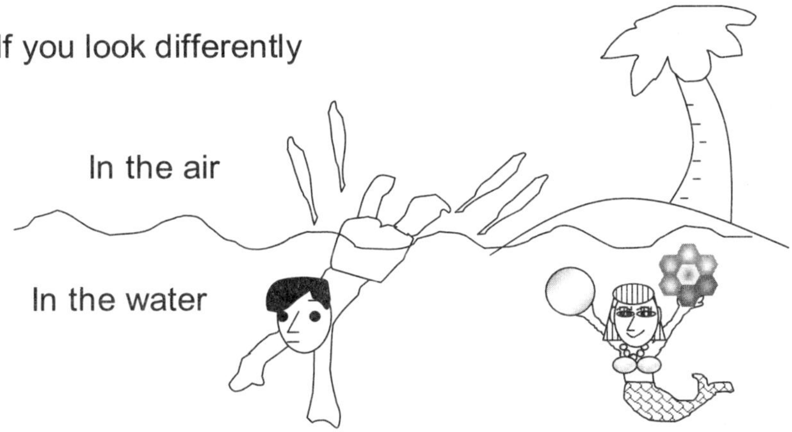

Fig.3 What is the internship on the whole?

For whom does a company exist?

For whom does a company exist? (Figure 1)

Let's start with this old and new question.

I have long believed that a company exists for its members (employees). That's why I work hard for my colleagues with whom I have shared good and bad times. In Japan, loyalty to companies comes from this belief. I was shocked to read the new Companies Act [12)], effective on May 1, 2006.

Since then, I have realized that a company exists for its shareholders (is it common sense?). The Companies Act defines shareholders as members of a company. We are just employees. Does a company exist for shareholders investing in the company or for employees working for the company? Of course, for shareholders.

In this era, the composition of shareholders may change overnight, followed by the change of the president and layoff of employees, no matter how hard you work with loyalty. I do not mean to discourage employees' loyalty, but we should work under a new company concept for our companies, for ourselves, and for our colleagues. Even when working for development-based companies, we must know what a company is. Let's learn the Companies Act briefly.

Companies Act and Financial Instruments and Exchange Act

What differences have been brought about by the two laws?

The Companies Act, effective on May 1, 2006, stipulates *internal control* to be observed by the boards of directors of large companies. The Financial Instruments and Exchange Act, effective on June 2006, requires that listed companies submit internal control reports from the fiscal year starting in April 2008. These laws are briefly explained below. Consult related books for details.

1. A company is financed by (1) debts from financial institutions (liabilities) and (2) equities (capitals) to make profits by investing (3) assets ((1) + (2)).
2. In the past, few shareholders (investors of (2)) demanded returns, but the composition of shareholders has recently changed. Recent shareholders demand returns (dividends) larger than interests that they would earn by depositing money in banks.
3. Shareholders need criteria to choose investees. While banks guarantee the *interest rate*, companies do not guarantee dividends.

Which financial instrument do you choose, a principal-guaranteed product with a low interest rate or a non-principal-guaranteed product with a high interest rate? If the latter is chosen, you will determine the risk by consulting with a securities company or by doing research by yourself.

Company stocks are such *high-risk* financial instruments.

Fig.1 Who owns this company?

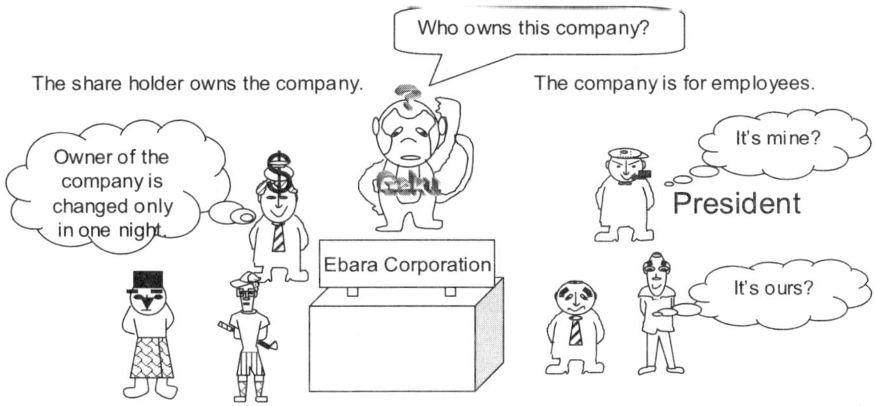

Fig.2 Company is one of Financial instrument

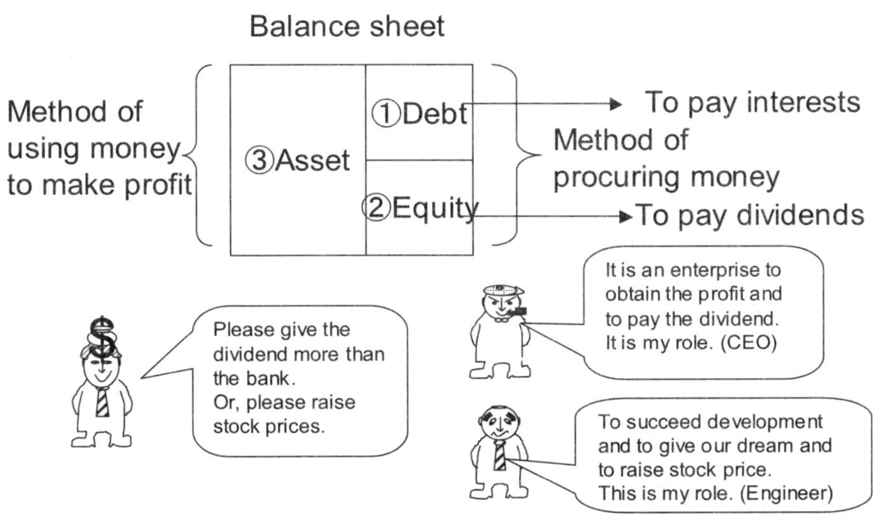

Handling patents requires legal knowledge.

Talking like a lawyer (Figure 1)

Development engineers need legal knowledge, as well as financial knowledge. In this era, engineers cannot conduct development without legal knowledge [13]. Let's learn legal knowledge that an engineer should have to talk *like a lawyer*. Be careful when using legal knowledge, as the proverb says, "A little knowledge is a dangerous thing."

Below you will find:

- Intellectual property related laws, which protect your ideas;
- Industrial Safety and Health Act, which ensures safe work;
- Product Liability Act, which assumes the possibility of accidents, and;
- Various environment-related laws.

The key points of these laws are outlined. Consult related books for details. You need legal knowledge to talk like a lawyer, not to be a lawyer.

Protecting your ideas: Intellectual property related laws (Figure 2)

Intellectual property related laws include the Patent Act, the Utility Model Act, the Design Act, and the Trademark Act. There are also more specific laws, such as the Act Concerning the Circuit Layout of Semiconductor Integrated Circuits to protect the circuit layout right for semiconductor integrated circuits.

The way to write patents will be discussed later. If you are an engineer, you have read a patent or written your own patent, haven't you?

The first-to-file and first-to-invent systems may be the first key point. Japan adopts the first-to-file system; patents are granted to the first party to file. If two or more parties come up with a similar invention, its patent is granted to the first party to file the application. The U.S. adopts the first-to-invent system; patents are granted to the first inventor. If you can prove that you are the first inventor, you are entitled to a patent even after another party files the application. There are difficult problems in the latter system, although details are omitted here.

Engineers must be careful not to publish their inventions at conferences or in papers before filing a patent application. When you have done so, remember that Article 30 of the Patent Act provides provisional exceptions if an application is filed within six months of the date on which novelty is lost by publication.

Engineers, intellectual property specialists, and patent attorneys must work closely to obtain strong patents (Figure 3).

Fig.1 R&D engineers should know financial affairs and laws.

Fig.2 How to file the patents

Fig.3 Firmly unites firmly.

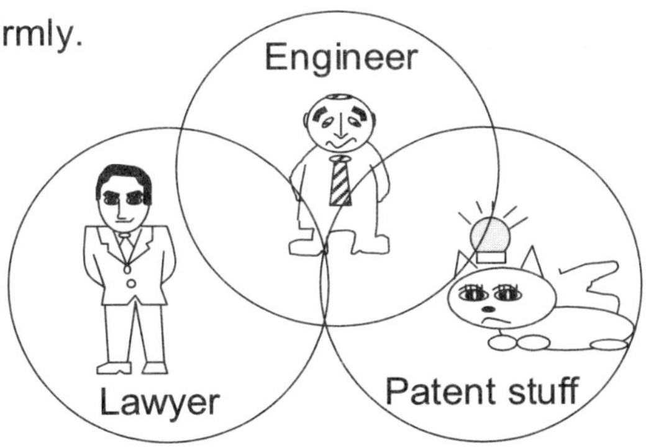

Accidents may occur anytime.

Product Liability Act (Figure 1)

The Product Liability (PL) Act took effect on July 1, 1995, when there was a concern that PL lawsuits would significantly increase as in the U.S. The concern turned out to be unfounded. Instead of fearing possible accidents, it is important for engineers to create reliable products and take sincere measures in the event of accidents.

When an accident is reported by a user, send a person closest to the accident site (generally a sales representative). Also, report the accident without delay to the designated personnel according to the emergency reporting system for risk management.

Then, start investigating the cause. Quality engineering estimates that 90 % of accidents are caused by flaws in design. Designers should sincerely respond to the accident with humility.

Following temporary measures, the quality control department takes permanent measures, including FTA (Fault Tree Analysis), based on quality engineering.

The legal department checks related laws/regulations to give advice, which is never sly one to stay away from lawsuits but one useful for understanding regulatory and legal frameworks correctly and for taking proper measures.

It is critical that the development leader, the division manager, and the president of the company express their will to "take all responsibilities."

Flying turbo pump (Figure 2)

Here is an example of accidents.

It happened in 1985 when we started the sales of a turbo molecular pump developed by Ebara. There were serious accidents; the pump units broke at many customer sites. The turbo molecular pump has blades that rotate at 50,000 rpm. While bending the system, the pump jumped out with a roaring sound. Anyone near the pump could have been injured.

The sales representatives and engineers were immediately sent to each site to replace the pump units, but the replaced ones also broke. We tried every possible measure based on quality engineering, but could find no solution. It was the reality of development. Struggling for six months exhausted the sales representatives and engineers. Finally, one of the engineers said, "Let us give up. We should withdraw all pump units and compensate for them."

It was a difficult decision. The division manager joined us to apologize the customer.

We felt sorry for the customer rather than frustrated.

I will never forget what the customer said.

"Mr. Tsujimura, it is easy to give up, but the development of semiconductor equipment is joint work between users and manufacturers. Before feeling sorry and apologizing, please find a solution."

Thanks to the customer's word, I have been able to continue the development of semiconductor equipment.

Fig.1 Product Liability

Fig.2 Flying TMP

Security and safety for our grandchildren

Working safely: Industrial Safety and Health Act (Figure 1)

The Industrial Safety and Health Act aims to prevent industrial accidents and ensure the safety and health of workers. It requires that officers responsible for the safety and health of workers be appointed. Supervising workers is also stipulated in the Act; the payment of compensation for accidents resulting from failure to observe these regulations has increased year by year. Of course, it is the best that no accident occurs. However, safety systems for ensuring the safety and health of workers have been established on the premise that accidents may happen anytime.

- Are workers enforced to work under poor work conditions?
- How is the safety of workers secured and confirmed?
- What measures are taken after an accident happened?

It is recommended to consider these points once.

For our grandchildren: Environment and laws (Figure 2)

The Basic Environment Law takes the central position in the structure of environment related laws. Article 21of the Law defines the following regulations, and it is advisable to at least know their titles. For details, consult related books.

1. Air Pollution Control Law, enacted in 1968.
2. Water Pollution Control Law, enacted in 1970.
3. Soil Contamination Countermeasures Law, enacted in 2002.
4. Offensive Odor Control Law, enacted in 1971.
5. Noise Regulation Law, enacted in 1968.
6. Act on Confirmation, etc. of Release Amounts of Specific Chemical Substances in the Environment and Promotion of Improvements to the Management Thereof
7. Basic Law for Establishing the Recycling-based Society, enacted in 2000.
8. Waste Management and Public Cleansing Law

Fig.1 In order to work safely

Fig.2 Environment is the most important

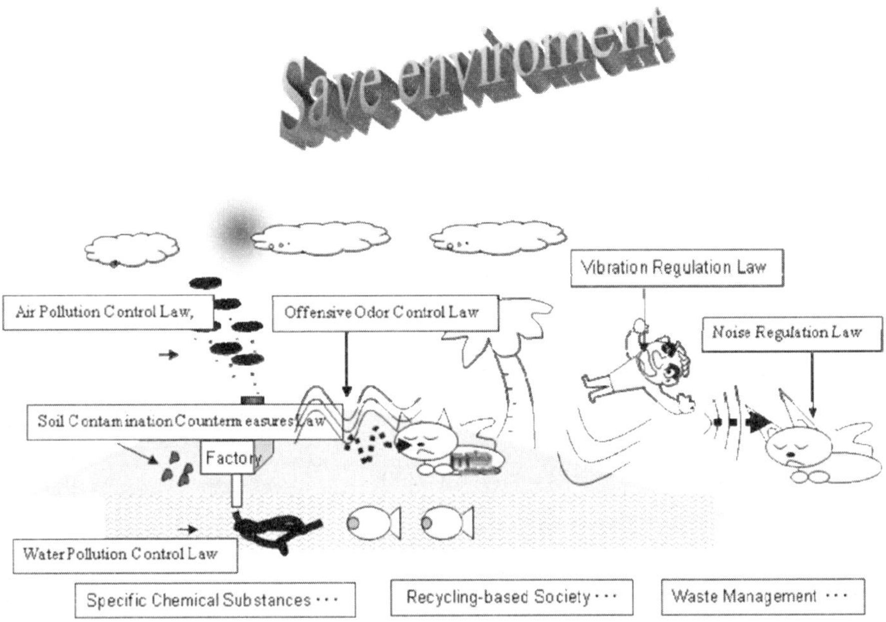

Elevator talk

I have applied a *15-second rule* in my workplace.
The rule is to finish oral reporting within 15 seconds.
"I finished XXX yesterday (5 seconds)."
"I am going to do YYY today (5 seconds)."
"I will start ZZZ tomorrow (5 seconds)."
15 seconds are enough for finishing oral reporting.
And 15 seconds are longer than you think.
Try keeping quiet for 15 seconds while looking at your watch.
"…………………………" (15 seconds)
You see? It is longer than you think.
I heard that TV commercials run for 13.5 seconds at the longest.
A longer commercial has an adverse effect to make viewers bored.
I have applied the 15-second rule for 20 years. When I talked about the rule abroad, I was told,
"Mr. Tsujimura, it's like an elevator talk. In the U.S., you will never be promoted unless you make good reporting when you see your boss or president in an elevator."
The same rule seems to apply everywhere.

In elevator

Chapter IV

R&D Born through Education

20 hours up to Clarkson university

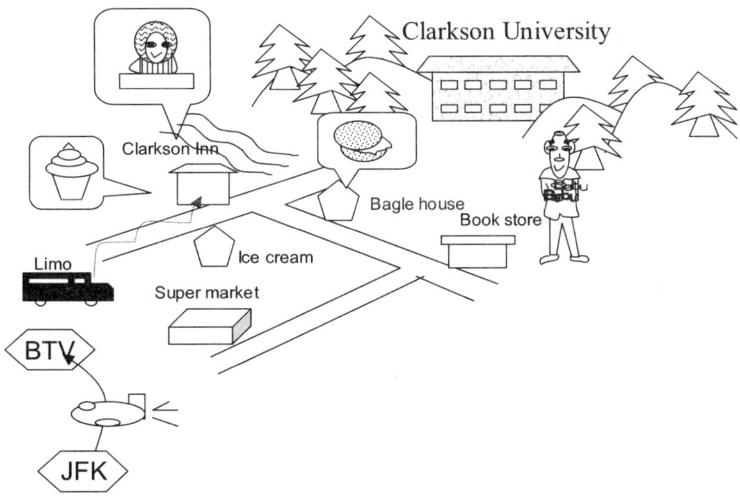

Professor, how can I get a PhD?

Company's order to get a PhD! (Figure 1)

One afternoon in 2000, I was called by my boss. I was busy but rushed into the boss's office.

"Mr. Tsujimura, you are the development manager of the first section listed company, but you don't have a PhD. You can get it easily, can't you? Go and get a PhD," said the boss.

I was surprised, actually. However, I was a fearless man, and I said,

"Yes, I understand. A piece of cake. Ha ha ha."

Looking back now, how foolish and disrespectful I was. Soon I was punished.

Hard enough to loose hair and teeth (Figure 2)

I watched a TV program featuring overseas students who came to Japan to get a PhD. A student made efforts with having a part-time job, but the student could not get a PhD in the end and returned to the home country. Another student could not get it for ten years. The hair was thinning, and teeth were lost. The student was exhausted both physically and mentally.

The TV program made me anxious little by little. I identified myself with the students on TV. In the TV program, an excellent academic advisor appeared, and the documentary ended with a happy ending.

Remembering that I had an excellent senior, I soon visited the university I graduated from to see my professor.

Two ways of getting a PhD (Figure 3)

The professor gently explained in an understandable manner as before.

According to his explanation, a PhD is the most prestigious honor awarded to individuals, and the professor is very proud of having it. After obtaining a PhD, I must devote myself to research without diminishing the pride.

That was the precondition for receiving instructions from the professor.

There were two ways. One was to get a PhD by writing theses, and the other was to enroll in a doctoral program for working people.

The former is literally the way to obtain the degree by writing a doctoral dissertation, while the latter is for working people to enter a doctoral program and obtain the degree, as is the case with students.

I selected the latter without any hesitation.

Fig.1 Get a PhD

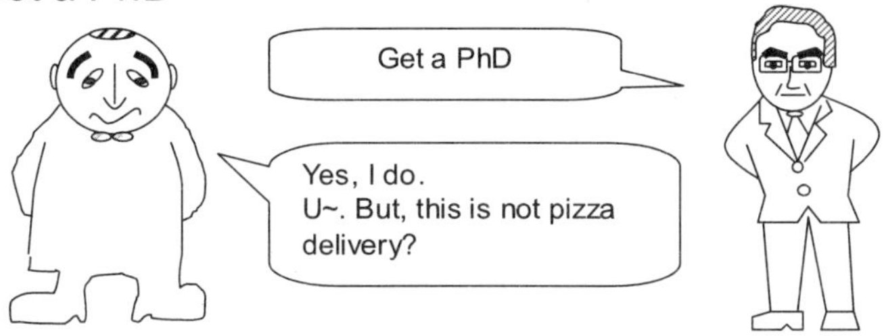

Fig.2 Loose my hair and teeth

Fig.3 There are two ways to get a PhD

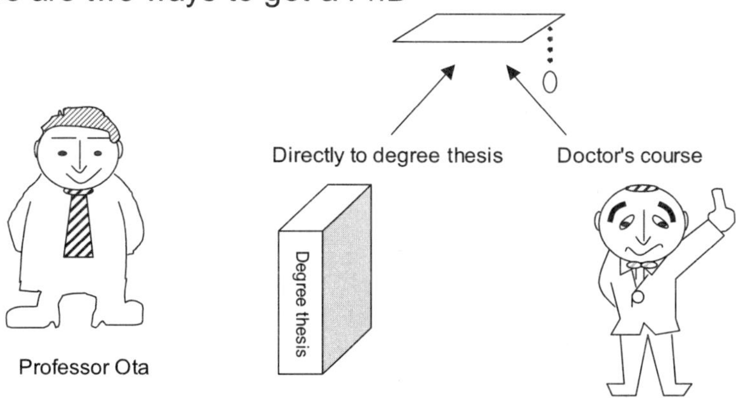

R&D and getting a PhD share the same road.

Interesting, but not enough (Figure 1)

First, I showed all of more than 100 publications and more than 100 patents I made over the past two decades to the professor.

"Mr. Tsujimura, your writings are very interesting, but they cannot be converted into papers," said my professor.

I was disappointed.

"But I'm surprised at such a large number of patents and at reliable data. I'm amazed at the good patents made by the leading company. Why don't you rewrite all of them into papers?"

At that moment, my belief about patents and papers was born.

Writing good patents facilitates writing papers.

Patents benefit companies, and papers benefit society.

Key point: storyline (Figure 2)

Next, I tried to build the storylines of theses.

As with dramas, a fundamental framework is important.

The professor instructed me on how to build storylines, but he never interfered with the contents.

It was his way of expressing affection.

He said that the will to write my own unique theses was important.

By recalling the past 20 years, I picked up various things, including what I wanted to do and write. In this way, I made more than 20 frameworks. The final product among them is shown in Figure 2.

[Background of development] The reason that the selected theme must be addressed was presented (marketing research was helpful).

[Key to development] Novelty and validity were emphasized (the basics of patent writing were applied).

[Approaches] The details of development projects I participated in were exactly reflected in this section. Discussions regarding necessary technologies and R&D activities for them led to the conclusion. For the doctoral dissertation, individual theses were made based on the respective elements.

[Conclusion] The point of objectives, i.e. the theme of the doctoral dissertation and the development projects, was shown.

Fig.1 It's interesting though

Fig.2 Story making is important. From my degree thesis

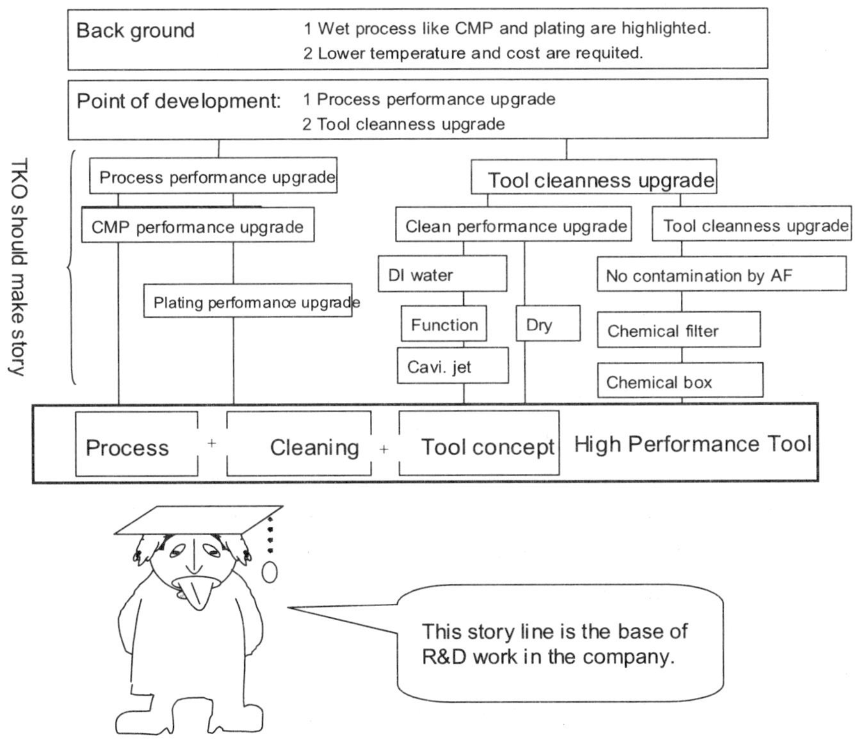

At the end of trials and tribulations

My encouragement to you to obtain a PhD (Figure 1)

The framework was approved by my professor.
I regained composure. I was relaxed and felt a desire.
I was nearly 50 year old (and was sure to turn 50 before I could get a PhD).
I tried to obtain a PhD because I sincerely wanted my young fellows to follow the same way.
I thought that, if even I could obtain it (I had not yet started writing theses then. I was hasty), everyone could do so. I decided to be a good predecessor.
I also wrote a recommendation paper entitled *You can get a PhD*.
Although it was easier to submit papers to one association only, I belonged to multiple associations and submitted papers to all of them, including:
The Japan Society of Mechanical Engineers (cavitation cleaning and structural dynamics of semiconductor multilayer interconnection);
The Japan Society for Precision Engineering (nanotopography and roll-off characteristics in CMP applications);
The Japan Society for Abrasive Technology (process analyses for CMP);
Japan Institute of Electronics Packaging (plating technology, including cap plating);
Material Research Society (MRS) (wet processes, etc.);
Advanced Metallization Conference (ADMETA) (multilayer technology), and;
International CMP Symposium (CAMP) (of course, CMP process).
All the associations provided paper templates. In addition to these templates, I noted down how to submit papers and how they were examined. Now, many young fellows often submit their papers by referring to the templates I used.

The day when I received the PhD diploma (Figure 2)

Finally, in 2002, the day of receiving my PhD diploma came. I joined the commencement ceremony with many students in their twenties. My 80-year old mother also attended the ceremony. It brought a tear to my eye, reminding me of the hard work.
Appreciating my mentor's help, I swore to lead a good life with the PhD.
It is great if the readers of this book are decided to try to get a PhD. I assure you that you will be happy as I am.

Fig.1 You can get a PhD too.

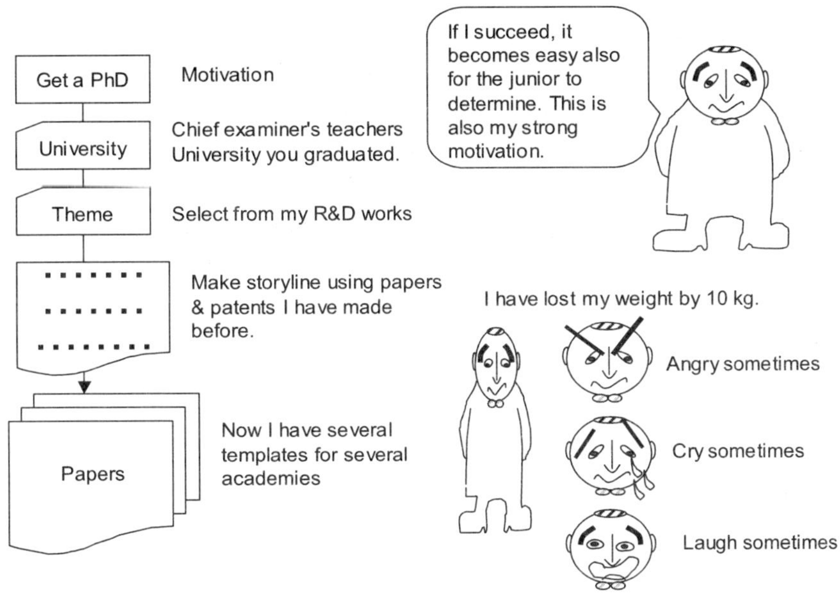

Fig.2 Thanks to all

Integrating theory and practice at Clarkson University (U.S.)

My recommendation to you to be a visiting professor abroad (Figure 1)

Getting a PhD will change your life in various ways. One of suggestions from my experience is to be a visiting professor overseas. The reason for being an overseas visiting professor is that overseas universities actively welcome visiting professors from companies. University professors cannot do what professionals at companies are doing, while universities are superior to companies in terms of theory. The integration of theory and practice can never be realized without cooperation between companies and universities. Keep in mind that the role of visiting professors is to bridge between them.

Recently, Japanese universities accept visiting professors (lecturers) from companies, and it is not necessary to adhere to going abroad. In the past, however, I received offers from overseas universities earlier. I will show you how I went to Clarkson University.

Encounter with Prof. Babu (Figure 2)

I have known Prof. Babu (former vice-president) of Clarkson University long before I obtained the PhD in 2002. From 1998 to 1999, we co-chaired the CMP symposium of the MRS meeting. Clarkson University has over a 100-year history and is traditionally well known for particulate research. Prof. Babu was already recognized as a foremost authority in the CMP field. Since I was working for a company having over a 100-year history and called myself a pioneer in the CMP engineering field, a rapport was soon established once we met.

He asked me to come to Clarkson University as a visiting professor immediately after I obtained a PhD. So, I contacted him as soon as I got it in 2002. From 2002 to date, I have been a visiting professor at the University.

There is an inside story. As a company employee, I must not be selfish. I asked my boss immediately. Fortunately, Clarkson University has produced many CMP engineers, most of whom are working for my client companies, such as IBM, Intel, and Micron. By explaining that working for the University would lead to the expansion of business, I received approval to go to the University during my summer vacation (Bon holidays).

You can become a visiting professor.
A chance to make a social contribution is prepared. Good luck.

Fig.1 Company's practice and university's theory

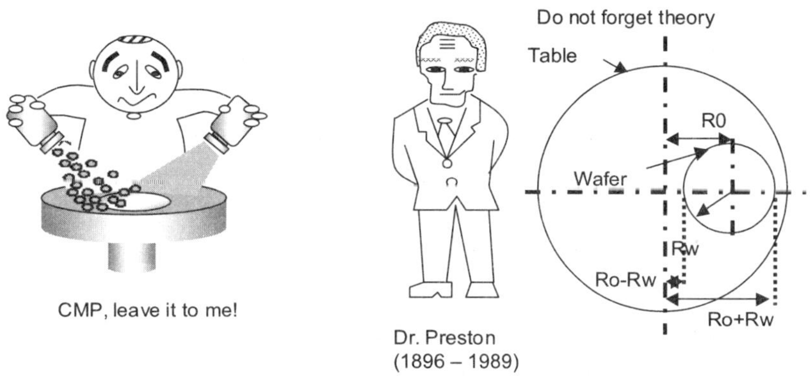

Fig.2 To Clarkson university

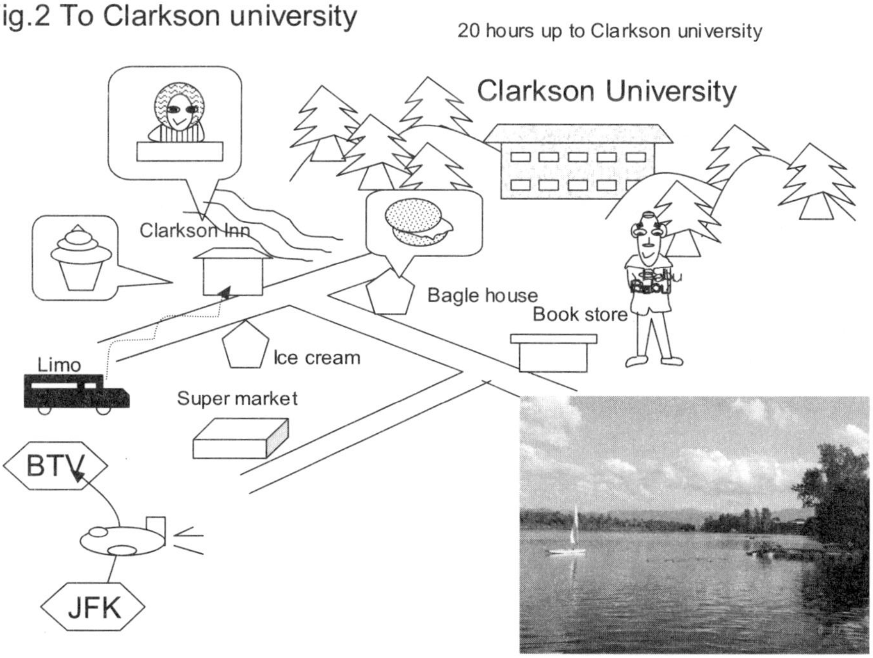

For the alumni of Clarkson University (U.S.)?

Tackling questions from students (Figure 1)

In 2002, I had my first lecture. I had much experience of making presentations at conferences, but no experience of giving lectures to students in a foreign language (English)! While a conference presentation lasts about 30 minutes, the lectures were half-day work with one class lasting 90 minutes.

Soon after finishing the dissertation, I started preparations for the lectures without taking a rest. Materials for ten lectures were prepared first, which would be the basis for *Handotai Wet Process Saizensen* (*Wet Process Revolution*).

The students at Clarkson University were very aggressive. They asked every kind of questions, and more than half of the time of my first lecture was spent answering questions. Actually, that situation was helpful for me. In addition, some questions revealed that my understanding of some issues was shallow. I put myself in the cycle of receiving questions and studying the topics. The interaction with the students helped me to learn more.

Clarkson alumni (Figure 2)

In 2008, it was the 7th time for me to give lectures at Clarkson University. I have had many occasions to communicate with various types of students. The University has CMP researchers from Asia, mainly from India and China; empathically, I cannot help but care their jobs. Most of them hope to migrate to the U.S. (some hope to go back to their home country). To do so, obtaining a green card (permanent resident status) is necessary. So far, I have written recommendation letters required to obtain a green card for many students. Every time the students graduate and report to me that they get the card, I am delighted and pleased as if I had a new child. This is my way to contribute to the alumni of Clarkson University.

At the CAMP (Figure 3)

Finally, I introduce CAMP [14], a CMP symposium hosted by Clarkson University and held for three days in Lake Placid (Winter Olympic games were held there twice) in August. The 13th CAMP was held in 2008. Contact me if you are interested in the symposium. Why don't you join discussions with CMP engineers around the world in Lake Placid in summer?

Fig.1 Every day of question attack

Fig.2 Clarkson alumni

Fig.3 CAMP (CMP academy)

Working for Tokyo Metropolitan University (Japan)

Working for Tokyo Metropolitan University

My mentor, Professor Ota of Tokyo Metropolitan University, knew that I was a visiting professor at Clarkson University (U.S.) since 2002. And I was given an opportunity to work for Tokyo Metropolitan University and became its visiting professor in 2004. My boss expressed understanding in this case again (actually, he may have thought that my appointment would be an advertisement for recruitment to the company).

Establishment of the three-step argument style (Figure 1)

For the first and second years at Clarkson University, I had no textbook in English. It was tough to prepare lecture materials for every class. In the third year and later, I had special classes instead of regular ones: a special lecture once a year at the University and a couple of presentations at the CAMP. Thus, I was considerably released from time constrains.

For visiting lectures at Tokyo Metropolitan University in 2004, I used the materials originally used at Clarkson University and translated from English to Japanese.

Visiting lectures in 2004 and 2005 exhausted me because I had lectures continuously from 10 am to 5 pm for three days. I was so serious that I followed through the lectures with no rest, and I was in bed with a fever afterward. Having cleverness in 2006, I developed the three-step argument style.

First, I let one student ask a question, but the question is typically vague. Second, I let another student repeat the question. Then, the intention of the question becomes clear. Finally, another student answers the question.

The development of the three-step argument style (question, confirmation, and answer) made me feel eased and was very beneficial for students. Hearing questions while thinking is really fruitful. Please try it yourself.

Publication of my book (Figure 2)

My first book, entitled *Handotai Wet Process Saizensen*, was made by compiling the lecture scripts. Focusing on frequently asked questions from students, the book has a reputation for being easy to understand. Since the original scripts were in English, it could also be published abroad, entitled *Wet Process Revolution*. I greatly appreciate the students' help.

Fig.1 Question, repeat and answer: Three-step argument style

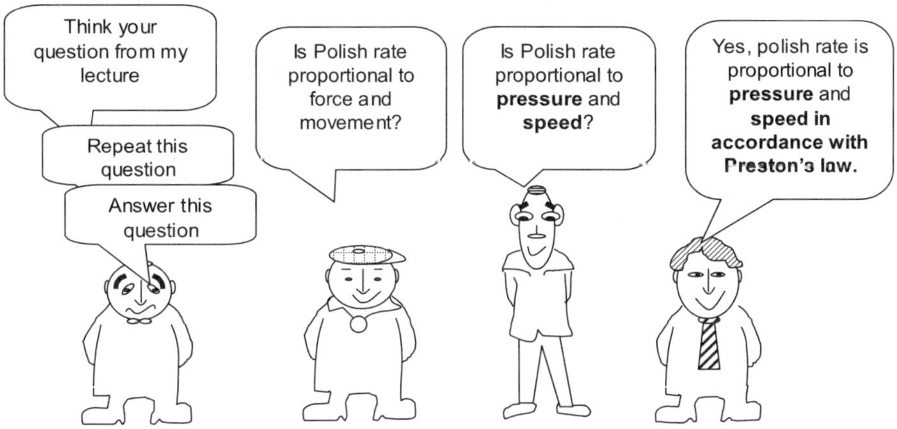

Fig.2 I published books

Working for Hanyang University (Korea)

Encounter with Prof. Park

Soon after becoming a visiting professor at Clarkson University, I received a mail for invitation from Prof. Park of Hanyang University in Korea. Hanyang University is a prestigious university that has produced many engineers in Korea. My friendship with Prof. Park has lasted since MRS Meeting 1998. I appreciated his offer, but could not accept it right then. At that time, I was already serving as a visiting professor at Clarkson University and Tokyo Metropolitan University. I hesitated to work in Korea additionally.

Holding the first international internship (Figure 1)

It was in 2005. Prof. Park made an interesting proposal. Hanyang University was planning international internships for students and selected Ebara as a candidate host company in Japan. Ebara is traditionally very active in social contributions and has experiences of arranging domestic internships. However, the company had no experience of arranging international internships.

I was ardent for it was the first time experience. I asked my boss immediately and got the go-ahead readily.

Then, the first international internship began with students from Korea. We had difficulties in the 2005 internship since it was the first time, and there was no infrastructure for international internships. It may have been difficult for the students, too.

Working in five countries and holding the Student Olympic (Figure 2)

A dream has grown up in my mind without my noticing. As described in "Preface" of this book, my dream is that, after serving as a visiting professor in five countries and retiring from the company, I establish a fund (from my modest retirement allowance) and invite students once a year to hold a convention entitled *The Evolution of Wet Revolution*. I appreciate your help in advance to make it possible.

Fig.1 Internship

Fig.2 From five countries in the world

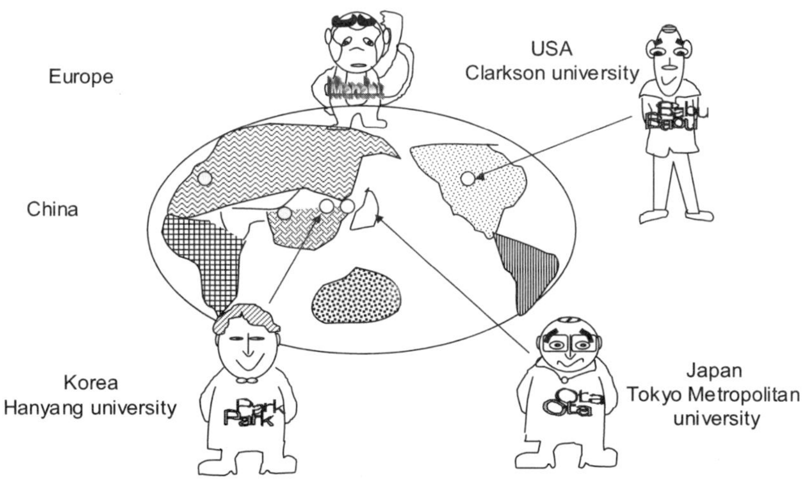

In languages in the world

Foreign languages as necessary accessories

As described earlier, I want to be a visiting professor in five countries; speaking in five languages is required. So far, I have studied Japanese (this is the most questionable), English, Korean, Indonesian, German, and Italian. Now, I am learning Chinese. I hope that this will suffice to realize my wish.

For reference, I will show you my foreign language skills.

Indonesian (Figure 1)

I was engaged in hydraulic power generation from 1974 to 1984 and spent most of the time in Indonesia. In deep mountains in Indonesia, people did not speak English. My Indonesian language skill was obtained through desperate efforts, and I compiled it into a book. This book is so excellent that reading it for two hours allows you to master general conversations in Indonesian. The book also features episodes of my life in Indonesia.

Remember at least "Apa kabar (how are you)" and "Terima kasih (thank you)."

Korean (Figure 2)

Being engaged in the semiconductor industry gives many chances to deepen relations with Korea. We naturally become familiar with the Korean language. Further, because of the popularity of the drama *Winter Sonata*, Japan had an unprecedented Korean boom. I came up with an idea from that phenomenon. If you like a Korean drama, you watch and listen to it in Korean. Listening in Korean, you will feel at least the atmosphere of the language. If you are not good at studying Korean, it is just because you do not get used to Hangul letters. So, if you are a fan of Korean dramas, I recommend you to read *Katakana de oboeru hoho* (Method to remember in Katakana), a book shown in Figure 2. This one has also gained a reputation.

Chinese (Figure 3)

Next, the Chinese language. Each Chinese word has four tones and gives different meanings according to the intonation. I created a dance named four-tone dance to facilitate remembering. I will leave it to your imagination as to whether the dance worked well.

Also, I have another method to master the Italian language.

(I've heard voices yelling, "That's enough!" Let's go on to the next topic.)

Fig.1 Indonesian language

Fig.2 Korean language

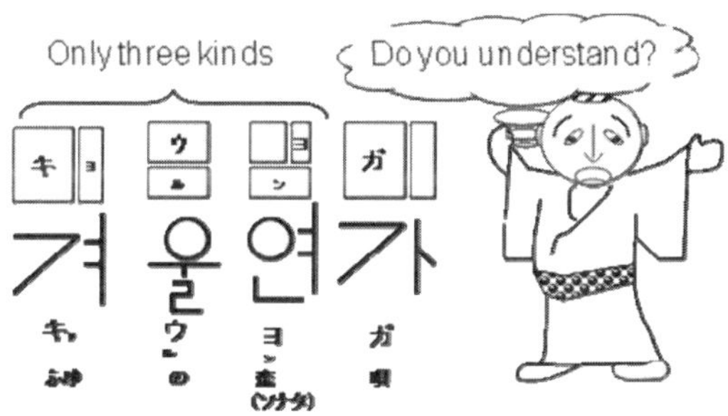

Fig.3 Chinese language

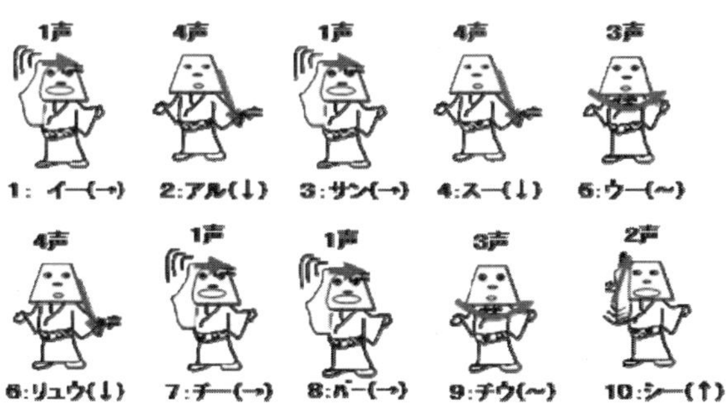

Knocked down in High Tech University!

Hello, high school students!

In recent years, the increase in the number of students who dislike science and technology has been a problem. In 2007, the Semiconductor Industry News raised a problem with an article entitled *Gijyutsu jinzai kuraishisu* (crisis in recruiting engineers). It pointed out that students were moving away from science and technology, making it difficult to secure human resources in the semiconductor industry. The article cited the following causes: (1) Negative effect of pressure-free education, (2) decrease in the number of teachers, (3) reporting policies of the media, and (4) working conditions for engineers.

The article said that this tendency was observed in all developed countries. The rate of students who wish to enter science and engineering departments is 5 % in the U.S., 12 % in Europe, 20 % in Japan, and 40 % in China. If this is true, teaching undergraduate students is too late. When thinking about the necessity of teaching high school students the fun and importance of science, I was given a great chance. That was High Tech University.

In 2007, High Tech University, hosted by Semiconductor Equipment & Materials International (SEMI), was held for high school students. The first meeting took place in Kumamoto Prefecture, and the second meeting in Ibaraki Prefecture.

Helplessly knocked down (Figure 2)

I have been responsible for R&D and newcomer education at Ebara for over 30 years and been a visiting professor at three universities for seven years. Even so, things did not go as I expected for the youths.

I started my lecture in anxious.

The lecture title was *Semiconductor Engineers Flying to the World.*

- Description of the nature of semiconductors
- Explanation on R&D
- How interesting it is to work in the semiconductor industry

Regardless of my desperate efforts, the students straightforwardly said that they could not understand what they could not understand.

I was messed up after the event.

I was helplessly knocked down!

I learned a lot of things from the event.

Fig.1 The science course seems to be disadvantageous.

Fig.2 Semiconductor Engineers Flying to the World.

JOY in High Tech University

Looking closely, observing, and acting (Figure 1)

Fortunately, I was selected as a lecturer again for the next year.

I was tense and prepared myself not to fail this time.

"Mr. Tsujimura, do not regard high school students as kids. Treat them as your company's employees and educate them freely as you want. They will prefer it," taught a high school teacher.

Thanks to the teacher's word, I saw the light and decided what to do.

In the lecture, I let the students act as Ebara's employees to understand my point of view. Employee education is my domain. First, the students learned the basics of R&D: "looking closely around them," "observing themselves," and "acting (strategies)."

Key point: Making judgment (Figure 2)

Next was training for self-judgment. Showing the picture in Figure 2, I let the students freely present their opinions with no condition.

A question "Is the old man in the middle you?" was very shocking.

When they understood that "a rabbit, turtle, and man have a race," I asked them whether the man overtook the rabbit or was overtaken by the turtle. Since the picture gives a wide range of possibilities, various opinions come out depending on how it is looked at.

I told the students that expressing opinions and making judgment were important.

JOY of R&D (Figure 3)

Finally, I talked about the main subject, R&D, to the students.

I used a word, JOY, to tell them to have joy in R&D and defy common wisdom (*Joshiki O Yaburo*). This method suggests that you recognize all sorts of common wisdom first and then defy them.

Let's defy common wisdom with high school students!

JOY!

This lecture has provided the foundation of Chapter II of this book and the concept of JOY.

I have still been struggling despite the past two experiences. Are you afraid of high school students?!

Fig.1 Invitation to employee during a day

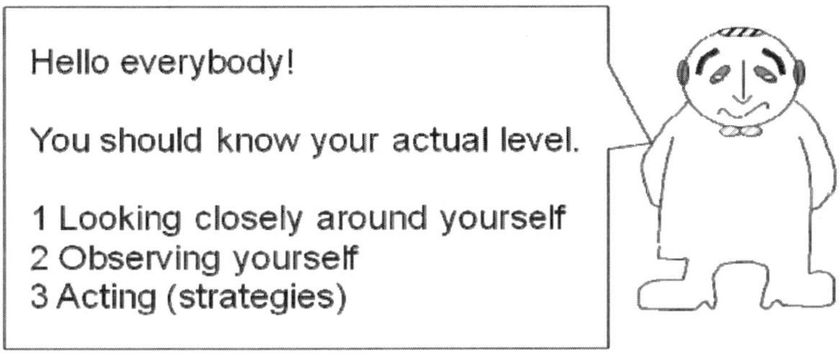

Fig2 Explain what happens from this picture.

There were a rabbit and a turtle in the park on the side of Boston and the Harvard University
I compete immediately with them. Which wins?

Fig.3 JOY

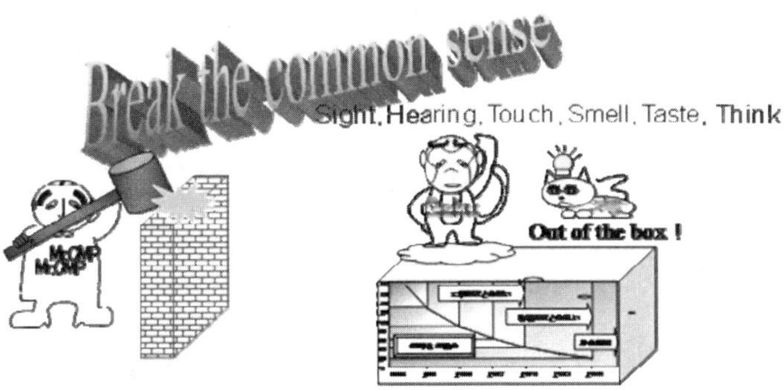

An elderly woman in Indonesia

It was a story in the 1970s, when I was working for hydraulic power generation in Indonesia.

At a village called Garung located in central of Java in Indonesia, a hydraulic power generation plant was under construction. The construction of the plant took about two years from the beginning to the opening ceremony. I visited the village many times. People in the middle of Java did not speak English (in those days), and I learned Indonesian in desperation. One day, I rode a local bus for a business trip. In the bus, a girl who looked like an elementary school student asked her grandmother again and again.

Girl: Gramma, what's that?

Grandmother: That's a cow.

Girl: Why is that cow wagging its tail?

Grandmother: Because the cow is whisking away flies.

Girl: For what?

Grandmother: Because, Tati (the girl's name), you don't like them flying around your face, do you?

Girl: Well, if flies don't go away, what is the cow gonna do?

I was impressed that the grandmother fluently and accurately answered the questions that seemed never to end. Since I started working on R&D later, I have often referred to the Indonesian elder woman. I think that, in R&D, if you persistently (always with fresh eyes like the girl) pursue Something New from the aspects of (1) Is it true?, (2) Why?, and (3) What is the mechanism?, the truth will be revealed.

Tati and her grandmother in Indonesia, Terima kasih (thank you).

Chapter V

How to Write Papers and Patents

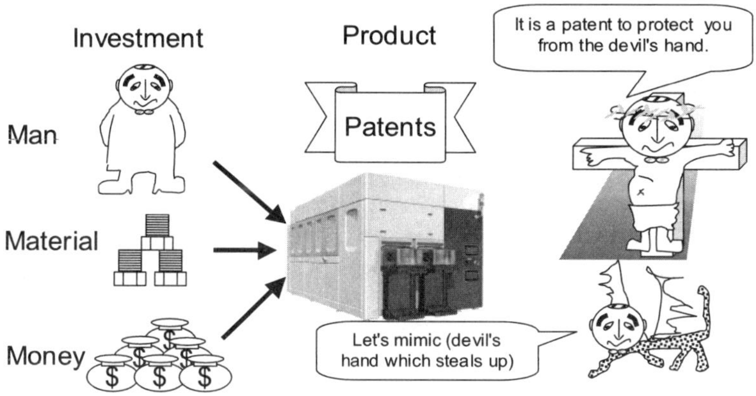

Patents benefit companies, and papers benefit society.

Patents for the benefit of companies? (Figure 1)

Development engineers at development-based companies are expected to undertake a critical mission to obtain patents. Patents protect your (or your company's) investments in development (people, goods, and money). They protect your (or your company's) rights for you (or your company). It is advisable to fully learn about patents and to master patent writing skills.

Patent writing is not difficult. Everyone can write patents by working on development in accordance with the five rules explained earlier.

Papers for the benefit of society? (Figure 2)

After patent publication, it is also advisable to write papers on the same subject. My experience shows that development projects based on definite ideas produce definite R&D data. Such data *greatly* facilitates the writing of papers.

The act of writing papers has various aspects.

Some may write papers for earning a degree.

Others may be advised by their bosses to write papers.

The most important purpose of writing papers is to disclose technology to the public and consequently to expand the circle of peers. Related peripheral technologies will be further developed, and breakthrough ideas may emerge. Anyway, there is no doubt that such technology advances with the efforts of peers. Thus, writing papers is a contribution to society, isn't it? It is indeed the primary contribution that development engineers can make to society.

***Lesson:** Write patents for the benefit of your company, and then write papers for contributing to society.

For your information, I have produced more than 100 patents and more than 100 papers. I have also received many offers for collaborative development based on my papers. This scheme may provide the Japanese version of Open Innovation (Figure 3).

Fig.1 You can defend the property by the patent.

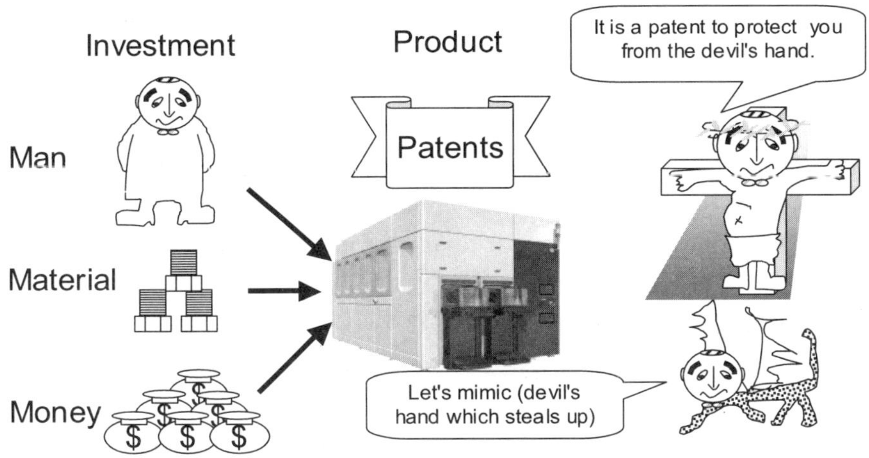

Fig.2 100 patents and 100 papers

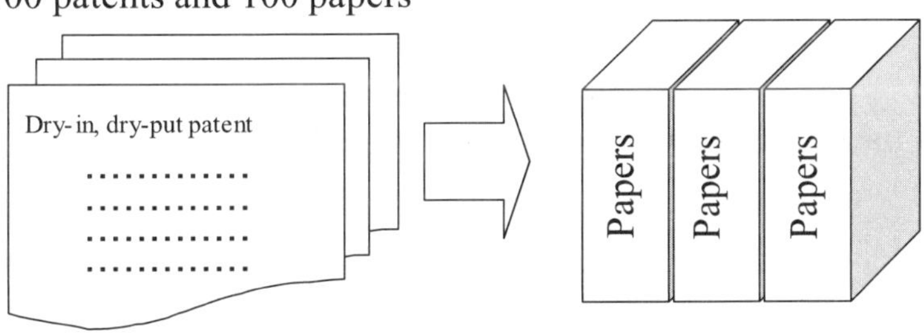

Fig.3 Circle of open innovation

Owning large tracts of wildland

Having more than 100 patents (Figure 1)

The other day, I met a staff member of the Japan Patent Office.

Mr. X (staff member of the Japan Patent Office): Japanese manufacturers should do more in the field of patent business.

Tsujimura: Are you sure, Mr. X? I have more than 100 patents, ha ha ha.

Mr. X: Mr. Tsujimura, suppose that you own large tracts of wildland in Hokkaido. Do you know how much tax you must pay? At present, there are numerous unused patents buried in companies or universities. Unused patents should be transferred to appropriate parties or revoked for yourself and society.

Oil-producing wildland? (Figure 2)

Tsujimura: But the wildland may produce oil (what a petty objection).

Mr. X: …

Some patents may be valuable like oil-producing land, but they are rarely among patents having value equivalent to that of wildland. Home runs (valuable patents) can be hit only by swinging for the fences.

Which is more important, quality or quantity of patents?

The conversation above implies several issues.

Of course, I believe that the quality of patents is more important than quantity. Patents having value equivalent to that of wildland are essentially unnecessary.

In the hydroelectric industry, to which I was assigned soon after joining Ebara, perfect patents would be submitted after full contemplation and adequate conceptualization.

That is not case with the semiconductor industry. I have often experienced that a delay of one day is critical in obtaining patents. Therefore, I changed my mind and decided to perform tests immediately after getting ideas and, if verified, to apply for patents without delay. In some cases, insufficient investigation or inadequate conceptualization spoiled my own possible patents. However, such cases are better than being behind competitors.

I intend to create high-quality and useful patents, often inspired by writing them.

For some time to come, I continuously apply for patents even if they are assessed to have value equivalent to that of wildland.

Fig.1 Patents like wildland

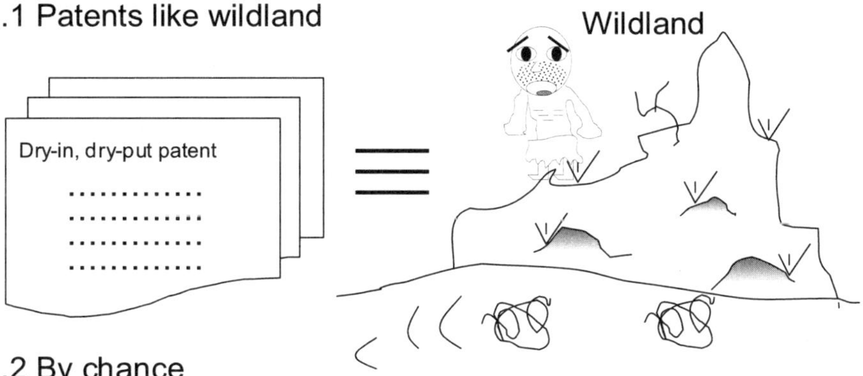

Fig.2 By chance

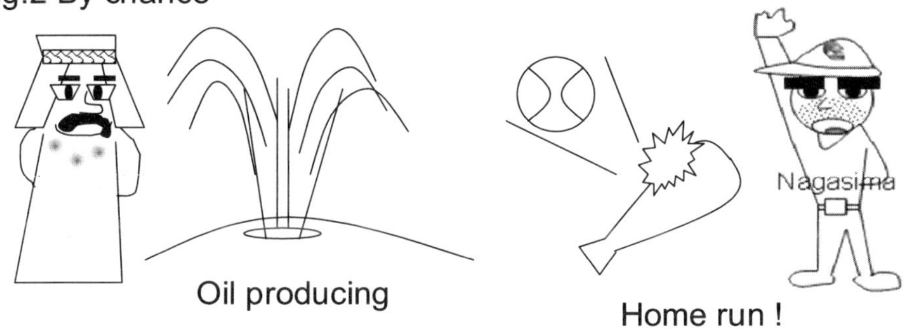

Fig.3 Various patents

How to write patents

The saying goes that "practice makes perfect."
Let's actually write patents.
Every company may have a form or template for patent writing. This section presents an easier method of writing patents (the method is introduced since it was actually used and well received by university and high school students).
Let's write a patent by following the instructions below in order.

What about prior art?

Aside from entirely novel ideas, most ideas come from inconvenience or problems associated with prior art technologies. Necessity is the mother of invention. Nonetheless, there is no way to invent a hat that does not interfere with angels' halo after going to heaven (Figure 1).

Describing the nature of the invention

The novelty and advantage of the invention must be described. First of all, explain what Something New is (novelty). However, an invention that is novel but has no benefit is useless; thus, it is also required to describe that the invention has the advantage of solving problems associated with prior art technologies. For example, a *table that allows eating food while standing on one's head* may be novel but is useless (Figure 2).

Presenting embodiments

Actually make drawings. The photos of fabricated items are also important.

Defining the scope of claims at the end of the patent

Pay extra care for writing the scope of claims since it is the effective part of the patent. In actual cases, advice may be available from patent attorneys and other specialists. Claims that define the invention more broadly should precede those that define the same more narrowly.

Case study (Figure 3)

First, figure out Something New. Recall and pick up the novel points of your idea. For your easy understanding, the example below explains how to write a patent relating to the dry vacuum pump mentioned earlier.
Prior art vacuum pumps use oil for sealing.
The nature of the invention is described.
(1) The pump according to the invention is novel in that it uses no oil.
(2) The pump according to the invention is advantageous in that it requires no oil, ensuring cleanness and safety.

Embodiments of the dry vacuum pump according to the invention are presented. The scope of claims is defined.

- Vacuum pump using no oil.
- Vacuum pump using non-contact sealing.
- Vacuum pump using shape-memory materials for non-contact sealing.

Fig.1 Hut you can use in the heaven

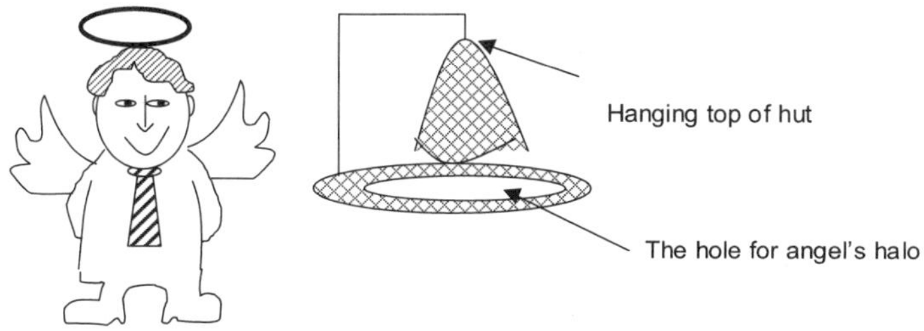

Fig.2 A table that allows eating food while standing on one's head

Fig.3 Start from widely, gradually narrow

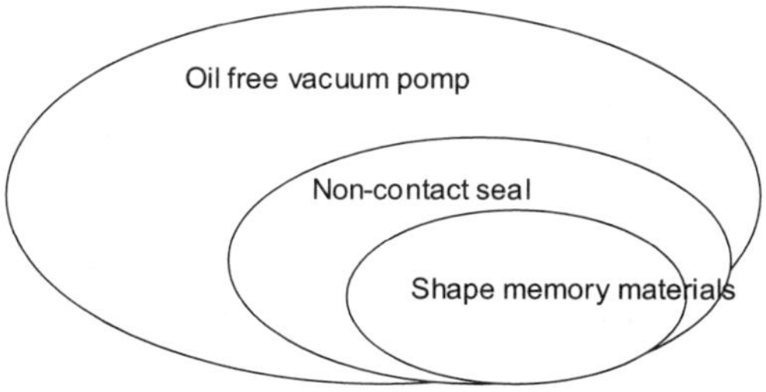

How about changing the size?

This section explains how to actually produce ideas. Only changing size specifications changes the way an invention is recognized. From this point of view, you may take numerous patents and be the world's No. 1 in any field.

Contest of strength (Figure 1)

Figure 1 shows a contest of strength between a man and an ant. Which do you think will win the contest?
You think that the man wins the contest, don't you?
However, what happens when the man shrinks to the size of ants?
Thus, the answer varies depending on how the scene is recognized.
Similarly, inventions involve various perspectives from above, below, the side, and every angle.

Tiny bubbles (Figure 2)

Let's look at an example of size change in the field of fluid engineering.

There are various types of bubbles, such as those in beer or generated by laundry machines. Do you know the size of the smallest bubbles that can exist?
I have seen bubbles of 1 mm.
I am afraid that I have never seen bubbles of 1 μm.
Then, can bubbles of 1 μm or less be formed?
I have consulted the *JSME Mechanical Engineers' Handbook: Fluids Engineering*. The Handbook does not cover such small bubbles. This fact means that an invention concerning the formation of such bubbles is patentable.
Hold on a minute.
Can bubbles of 1 μm exist? If so, what is their principle or mechanism?
And, what do such bubbles serve for?
A patent can be created by answering these questions in order (Figure 3).

Note) The development of ink-jet technology has been based on the study of micro-bubbles as mentioned above. In recent years, micro- and nano-bubbles are actively studied. Accordingly, the handbook at present contains the specifications for them. The example above was extracted from a conversation I had with a faculty member of Tokyo Metropolitan University before 1990.

Fig.1 Which is more powerful ?

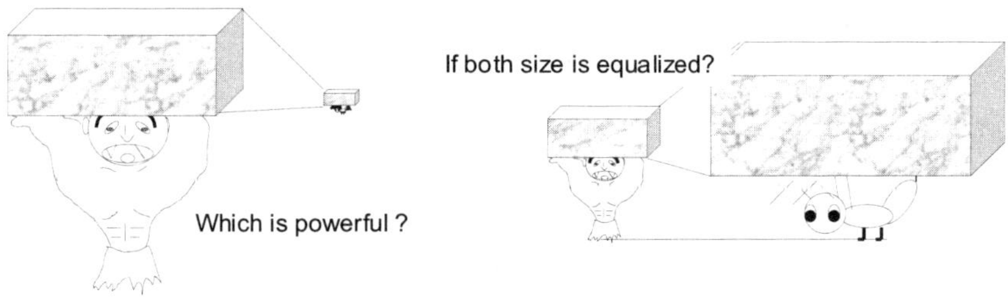

Fig.2 If size is changed?

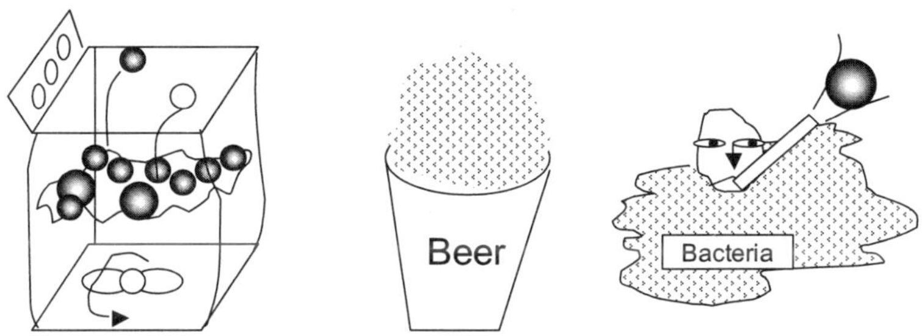

Bubble in washing machine: cm

Bubble in beer: mm

Soap bubbles made by bacteria: μm

Fig.3 Something new is important in R&D

How about changing the field of application?

In addition to size changes, a look should be taken at different fields. Only changing the field of application changes the way an invention is recognized. From this point of view, you may also take numerous patents and be the world's No. 1 in any field.

Which is a stranger? (Figure 1)

Do you know the movie *Planet of the Apes*? Featuring Charlton Heston, the film tells a story of astronauts crash-landing on a planet ruled by apes (the details of the movie are omitted here). In the movie, a scientist ape saw one of the astronauts (Charlton Heston) and uttered that perhaps this human (as if talking about an ape) might be a mutant that understand what we say.

The impact of human beings being replaced by apes corresponds to that of changing the field of application. Thus, a technology already applied in an area may provide patentable novelty in other areas.

Borrowed technology for semiconductor processing (Figure 2)

In the past, I learned the following from an engineer of a semiconductor device manufacturer: Most semiconductor processes are based on borrowed technologies, and existing technologies should be learned and used if appropriate.

Similar to the semiconductor industry, mature technologies are also applied in the highest-tech industries. Then, how about applying them more ambitiously?

A good example is the application of plating and polishing to semiconductor manufacturing. Copper plating dates back to ancient times; silicon oxide (stone) polishing is a mature technique. They were introduced to the industry by predecessors like the astronauts warping to the Planet of the Apes.

Note) Many were at first opposed to introducing CMP (polishing) to semiconductor processing. They refused to use CMP, saying "How dirty it is!"

Application to biotechnology and MEMS (Figure 3)

In the future, semiconductor technology may be applied to MEMS, NEMS, and biotechnology. Efforts must be made to ensure semiconductor technology is sufficiently mature before its application to these future technologies.

Fig.1 If field is changed?

Fig.2 Borrowed things from maturity field

Fig.3 To the next generation

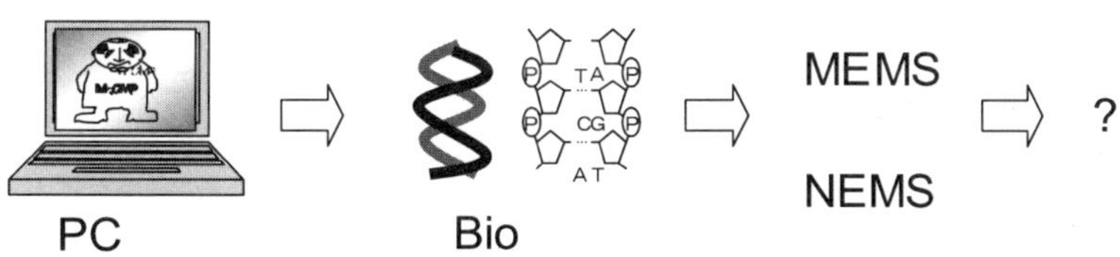

How to write papers

The saying that "practice makes perfect" is also applicable to paper writing.
First, let's actually write papers.
Forms or templates for paper writing are available from the websites of all academic societies. This section presents an easier method of writing papers (the method is introduced since it was actually used and well received by university and high school students).

A simple example is given below. Let's write a paper by following the instructions below in order.

Paper title: Damascene structure analysis using the finite element method (FEM) [15)]

Introduction: Provide the purpose and abstract of the paper.

The progress of scaling has led to the adoption of low-k materials with low structural strength for damascene structures, causing the problem of wafer breakage during polishing. To address this problem, the allowable limit of polishing pressure was estimated through experiments and analysis.

Data:

For three types of damascene modules, their respective stresses were analyzed using FEM. Then, experiments were conducted to identify the range of polishing pressure that actually causes breakage, and their results were compared with the analysis results.

Discussions:

A safety factor of two was observed for the 140 nm generation, while 0.5 was observed for the 45 nm generation.

Conclusions:

Summarize the data and discussion in a list form.

Acknowledgement: The author(s) would like to thank Mr./Ms. XX.

References:

Attached a list of references used for the paper.

Fig.1 Example of paper

A structural analysis at micron level was executed by the finite element method, and the effectiveness of the finite element method was shown including the experiment.

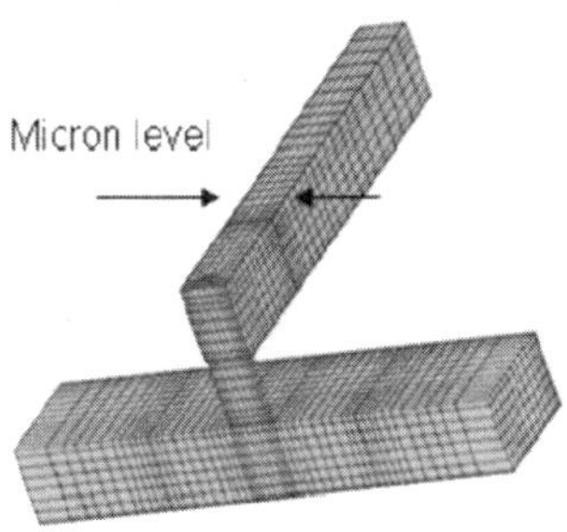

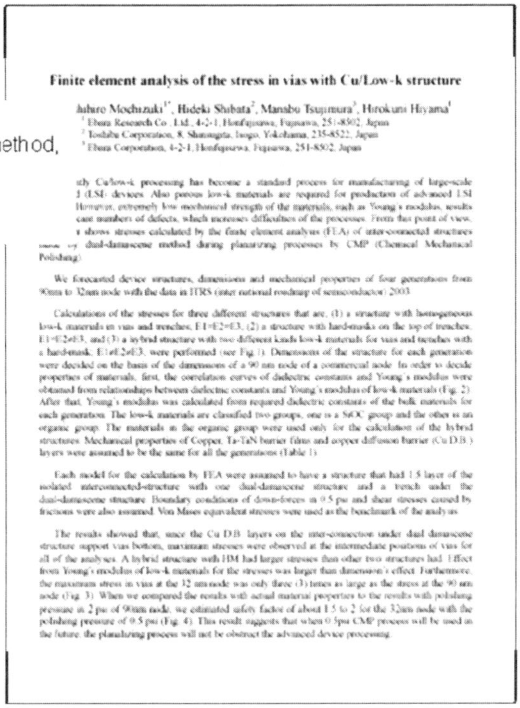

Finite element analysis of the stress in vias with Cu/Low-k structure

...ihiro Mochizuki[1*], Hideki Shibata[2], Manabu Tsujimura[3], Hirokuni Hiyama[1]

[1] Ebara Research Co., Ltd., 4-2-1, Honfujisawa, Fujisawa, 251-8502, Japan
[2] Toshiba Corporation, 8, Shinsugita, Isogo, Yokohama, 235-8522, Japan
[3] Ebara Corporation, 4-2-1, Honfujisawa, Fujisawa, 251-8502, Japan

...tly Cu/low-k processing has become a standard process for manufacturing of large-scale ...I (LSI) devices. Also porous low-k materials are required for production of advanced LSI. However, extremely low mechanical strength of the materials, such as Young's modulus, results ...cant numbers of defects, which increases difficulties of the processes. From this point of view, ...s shows stresses calculated by the finite element analysis (FEA) of inter-connected structures ... by dual-damascene method during planarizing processes by CMP (Chemical Mechanical Polishing).

We forecasted device structures, dimensions and mechanical properties of four generations from 90nm to 32nm node with the data in ITRS (inter national roadmap of semiconductor) 2003.

Calculations of the stresses for three different structures that are, (1) a structure with homogeneous low-k materials in vias and trenches, E1=E2=E3, (2) a structure with hard-masks on the top of trenches, E1=E2≠E3, and (3) a hybrid structure with two different kinds low-k materials for vias and trenches with a hard-mask, E1≠E2≠E3, were performed (see Fig.1). Dimensions of the structure for each generation were decided on the basis of the dimensions of a 90 nm node of a commercial node. In order to decide properties of materials, first, the correlation curves of dielectric constants and Young's modulus were obtained from relationships between dielectric constants and Young's modulus of low-k materials (Fig. 2). After that, Young's modulus was calculated from required dielectric constants of the bulk materials for each generation. The low-k materials are classified two groups, one is a SiOC group and the other is an organic group. The materials in the organic group were used only for the calculation of the hybrid structures. Mechanical properties of Copper, Ta-TaN barrier films and copper diffusion barrier (Cu D.B.) layers were assumed to be the same for all the generations (Table 1).

Each model for the calculation by FEA were assumed to have a structure that had 1.5 layer of the isolated interconnected-structure with one dual-damascene structure and a trench under the dual-damascene structure. Boundary conditions of down-forces in 0.5 psi and shear stresses caused by frictions were also assumed. Von Mises equivalent stresses were used as the benchmark of the analysis.

The results showed that, since the Cu D.B. layers on the inter-connection under dual damascene structure support vias bottom, maximum stresses were observed at the intermediate positions of vias for all of the analyses. A hybrid structure with HM had larger stresses than other two structures had. Effect from Young's modulus of low-k materials for the stresses was larger than dimension's effect. Furthermore, the maximum stress in vias at the 32 nm node was only three (3) times as large as the stress at the 90 nm node (Fig. 3). When we compared the results with actual material properties to the results with polishing pressure in 2 psi of 90nm node, we estimated safety factor of about 1.5 to 2 for the 32nm node with the polishing pressure of 0.5 psi (Fig. 4). This result suggests that when 0.5psi CMP process will be used in the future, the planalizing process will not be obstruct the advanced device processing.

Fig.2 What are points in this paper?

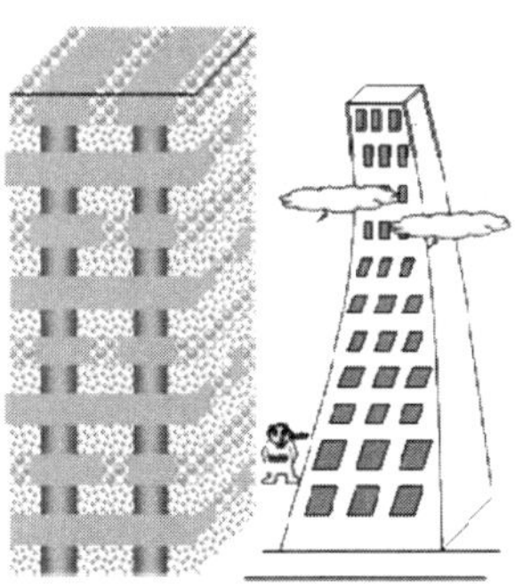

InterconnectSkyscraper

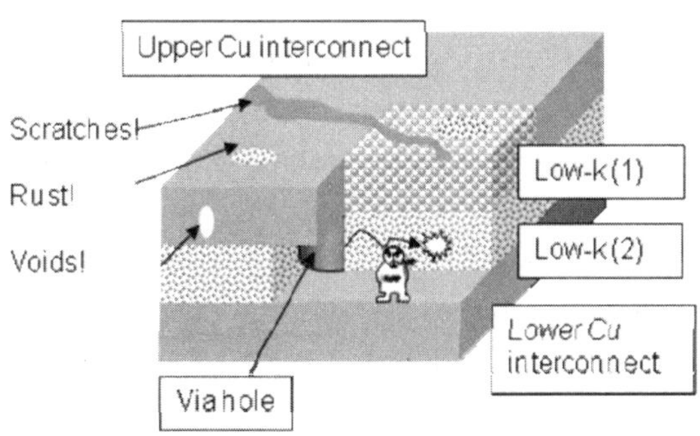

Let's start ACE!

Business persons good at reporting

Business people reluctant to write papers may be excellent at reporting or presenting their work with PowerPoint. An idea hit me. They should first be trained on how to write case study reports toward acquiring paper writing skills.

What are case study reports?

Yamamoto, who joined a company this year, is engaged in testing for metal polishing. There are many complaints that finished devices have numerous scratches. Yamamoto is writing a report for submission to his boss.
Phenomenon: The device surface was scratched by metal polishing.
Possible cause: The cause of scratching was estimated using FTA. Scratching may be caused by the following two possible factors:
Factor 1: Falling of diamonds from the dresser, and;
Factor 2: Falling of abrasives adhering to the unit's wall.
Demonstration: Tests were conducted for Factors 1 and 2, and the phenomenon could be reproduced.
Measures:
For Factor 1, a dresser with diamonds resistant to falling was adopted.
For Factor 2, the frequency of cleaning the unit's wall was increased.
Conclusions: The measures proved satisfactory with tests on an actual system.
The structure of the report is as described above.

Technical papers and case study reports

For technical papers, brainwork is required to generalize phenomena, identify mechanisms, and formulate equations, in addition to reporting facts. It can be aided by universities. Companies have a lot of topics and test data for papers, while universities can generalize such topics/data. This perspective provides a basis for collaboration between companies and universities.

Let's start ACE!

Ebara has a scheme for young engineers to be familiar with writing papers toward earning a degree. Engineers good at writing reports first write case study reports. The reports are then passed on to universities and finished as technical papers. This scheme is very effective. I have named the internal academic scheme the Advanced Conference of Ebara (ACE) on my own.

Reports to the superior
with the power point.
You are indeed good at this.

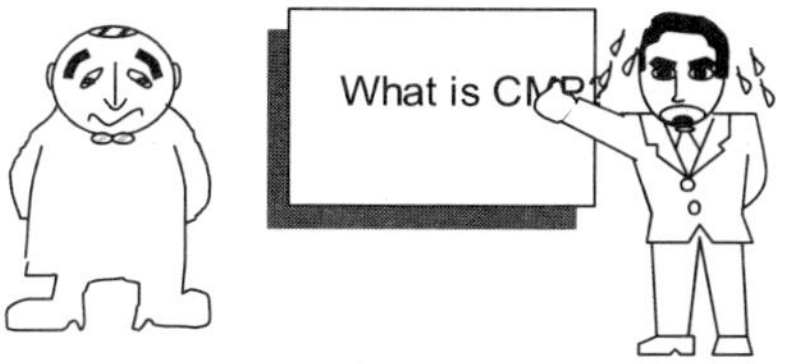

Case report

The content is a mere report
though it is a thesis form.
First of all, it is necessary to be
able to become a thesis form.

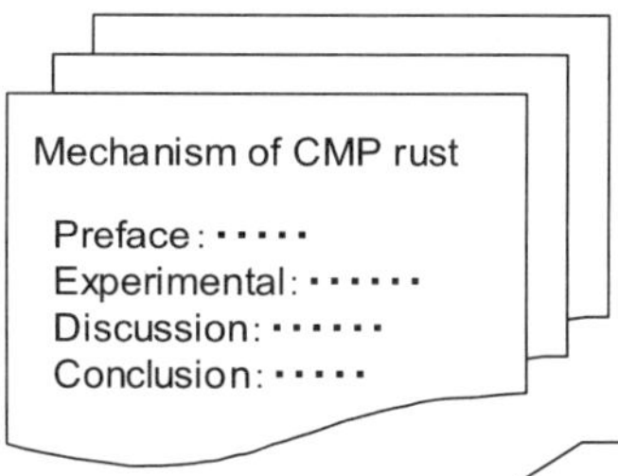

Generalization including theory

However, let's put the theory at the university.
Because, there are a lot of data.

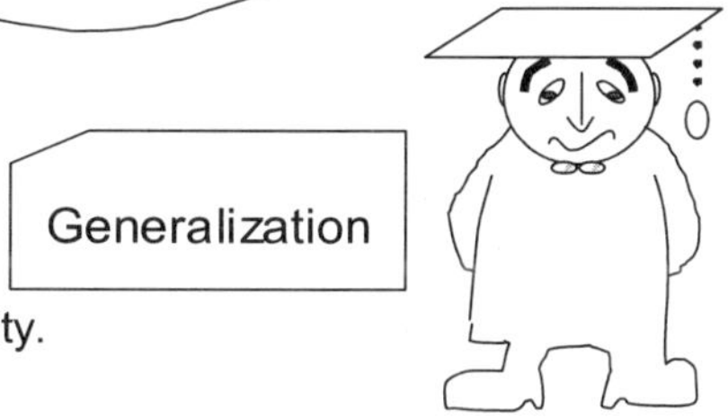

Thesis completion

Let's submit it to the academy with referee.

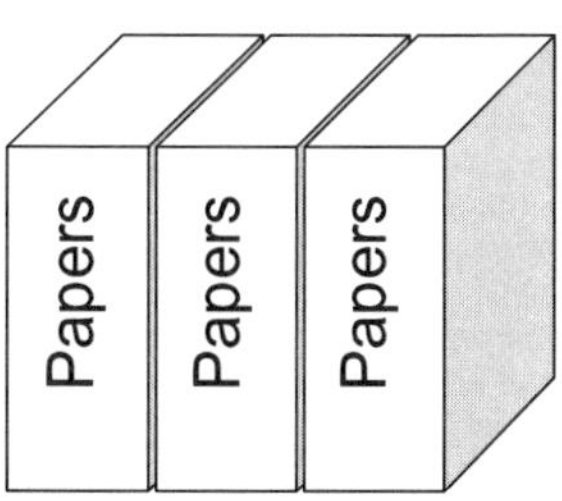

Patent trolls

This is a scary story.
Patents are, in principle, intended to protect the rights of inventors. However, they can be seen from a different angle. What happens when patents are taken, but products based on them are not released?
It is disadvantageous to the entire human community. Such patents are not justified.
In the semiconductor device industry, cross licensing is available. This method emphasizes the release of best products without taking care of patents (in fact, with some care taken for them). When patents are infringed as a result, the method assumes that cross-license agreements (mutual licensing) are concluded and that the gap in interests, if any, is compensated for in money.
There is no problem if this method fits all situations. However, a terrible problem occurs when some parties withdraw from the industry. One party to a cross-license agreement cancels the agreement, and the balanced relationship breaks. The financial burden on the other party suddenly increases.
Furthermore, a business model has emerged in which only patents are purchased from withdrawing parties and sold to winners remaining in the industry.
These businesses are called patent trolls (fearsome creatures).
It is really sad that *unscrupulous* people are utilizing patents as financial instruments, not for manufacturing. I hope that a hero will emerge to fight against these trolls.

Chapter VI

Five Revolutionary Technologies

The end of the world or the coming of a savior?

The end of the world? (Figure 1)

The 21st century has seen many issues that make us fear for the end of the world, such as population explosion, depletion of the Earth's resources (petroleum, etc.), and global warming. In July 2008, the Hokkaido Toyako Summit was held in Japan to seriously discuss ways to combat global warming.

The Cool Earth initiative, led by Japan, involves the creation of cool devices. In this light, the semiconductor industry has a mission to save the world from doom. All stakeholders around the world, including the global community, Japan, companies, and individuals, should make efforts to save the Earth from their respective standpoints.

What is semiconductor warming? (Figure 2)

An Inconvenient Truth by former U.S. Vice President Al Gore is famous for warning the dangers of global warming. A similar heat problem is present in the field of semiconductor devices. In the past, with the scaling (higher integration) of bipolar devices, the current density level was about to exceed 12 W/cm^2. This problem was solved by the emergence of CMOS devices, which operate at significantly lower current densities (the current density level is equivalent to the heat generation rate). While CMOS devices are approaching their limits in recent years, the concept of cool devices based on multi-core technology is emerging. The semiconductor device industry has not yet faced an inconvenient truth.

This situation will favor companies that manage many servers in data centers. By adopting low-power servers and lighting (LED, etc.) as well as power supplies from solar cells or hydraulic power generation, such companies can become a model for green IT companies.

Let's look into the development of cool devices in the following sections.

Fig.1 The end of the world?

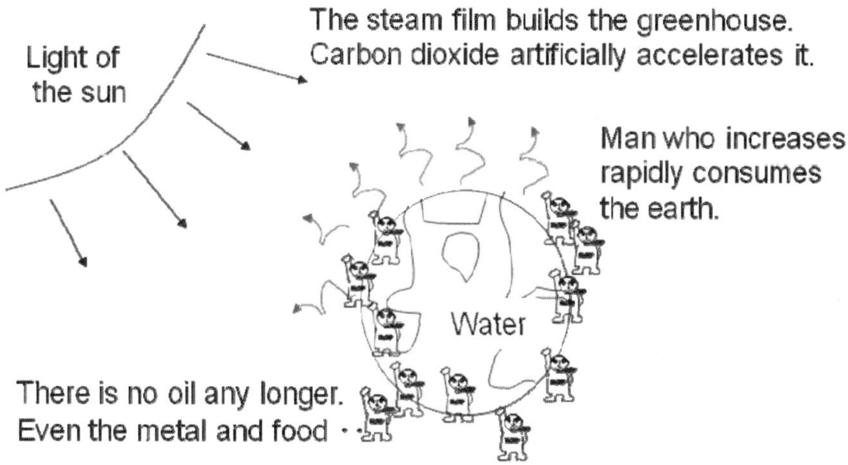

Fig.2 The IT technology will save the world.

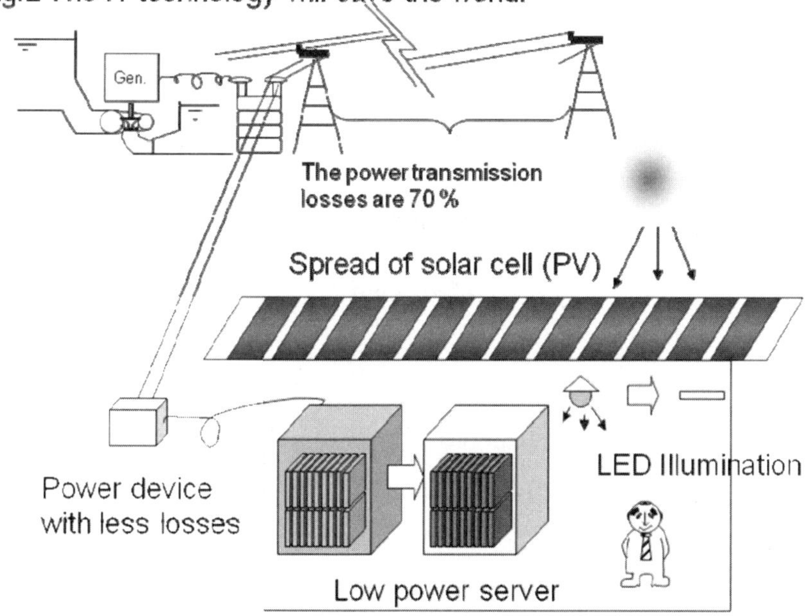

Descent of the Five Gods

The end of cool device development? (Figure 1)

Actually, multi-core technology is the result of compromise. It can be said that there was no choice but to modify the device design and operation since further scaling was difficult, and no novel device was conceived. The practical development of multi-core devices is just beginning. Facing both engineering and physical limits, the semiconductor device industry is currently expected to perform godlike miracles.

Help from the Five Gods (Figure 2)

When I felt discouraged and concentrated on praying, five ideas hit me. They are quite miraculous, and I believe that they are gifts from the Five Gods (?).

The five ideas are described below.

First, current issues in semiconductor processing and multilayer interconnection are discussed.

Planarization at once => **God with Divine Hands**

Fast planarization => **Superfast God**

Dummy insertion => **God with Metabolic Syndrome**

High cost efficiency => **Porcupine God**

Terrifying! => **Violent God**

Fig.1 Scaling level finally enter into God's region

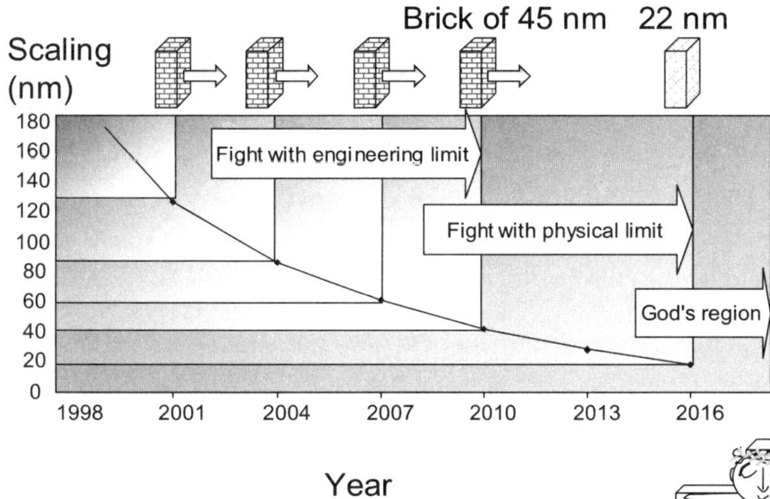

Fig.2 Five Gods

Paradigm Shift 45: Raising troublesome issues

Paradigm Shift 45

Some say that scaling has no future. In the view of "No More Moore," they argue that semiconductor devices no longer evolve with scaling. Others say that scaling is still promising in the view of "No! More Moore." There are also people who devise other different approaches in the view of "More than Moore."

This fierce controversy has resulted from the following circumstances. For more than 50 years after the invention of semiconductor devices, the device industry has enjoyed prosperity by taking advantage of scaling and wafer size transitions. Currently, the industry seems to be on the verge of a paradigm shift and has encountered troublesome issues related to the number *45*. The author has called this phenomenon *Paradigm Shift 45* and presented it since 2005 [16)].

A great paradigm shift is inevitably accompanied by the creation of revolutionary technologies. Current technical challenges are outlined below.

Scaling ends at 45 nm, and then…

Scaling is subject to both engineering and physical limits.

Engineering limit: For the 45 nm generation and later, particles, which cannot be detected optically, increase exponentially, and they are difficult to remove by van der Waals forces. In engineering terms, extreme difficulties must be overcome.

Physical limit: For the 45 nm generation and later, the resistance of Cu interconnects increases with the scattering effect. In physical terms, the industry reaches a level at which physical constants are affected.

3D integration is implemented with 45 μm, and then…

3D integration is in the spotlight, but thinner wafers require significant changes in wafer processing technology. Unlike wafers of about 100 μm, those of 45 μm or less are difficult to handle.

The wafer size transits to 45 cm, and then…

The wafer size transition increases radial non-uniformity depending on the process type. Typically, since it coincides with scaling, development engineers face two difficulties: tougher specifications with scaling and more severe requirements for radial uniformity with the wafer size transition. The cost of developing 45 cm wafers and related systems is huge, making us reluctant to introduce 45 cm wafers. A wafer size of 45 cm is subject to economic limits.

Paradigm Shift 45, which I have proposed since 2005, is as explained above. The paradigm shift may take us to a paradise or bring about the end of the world.

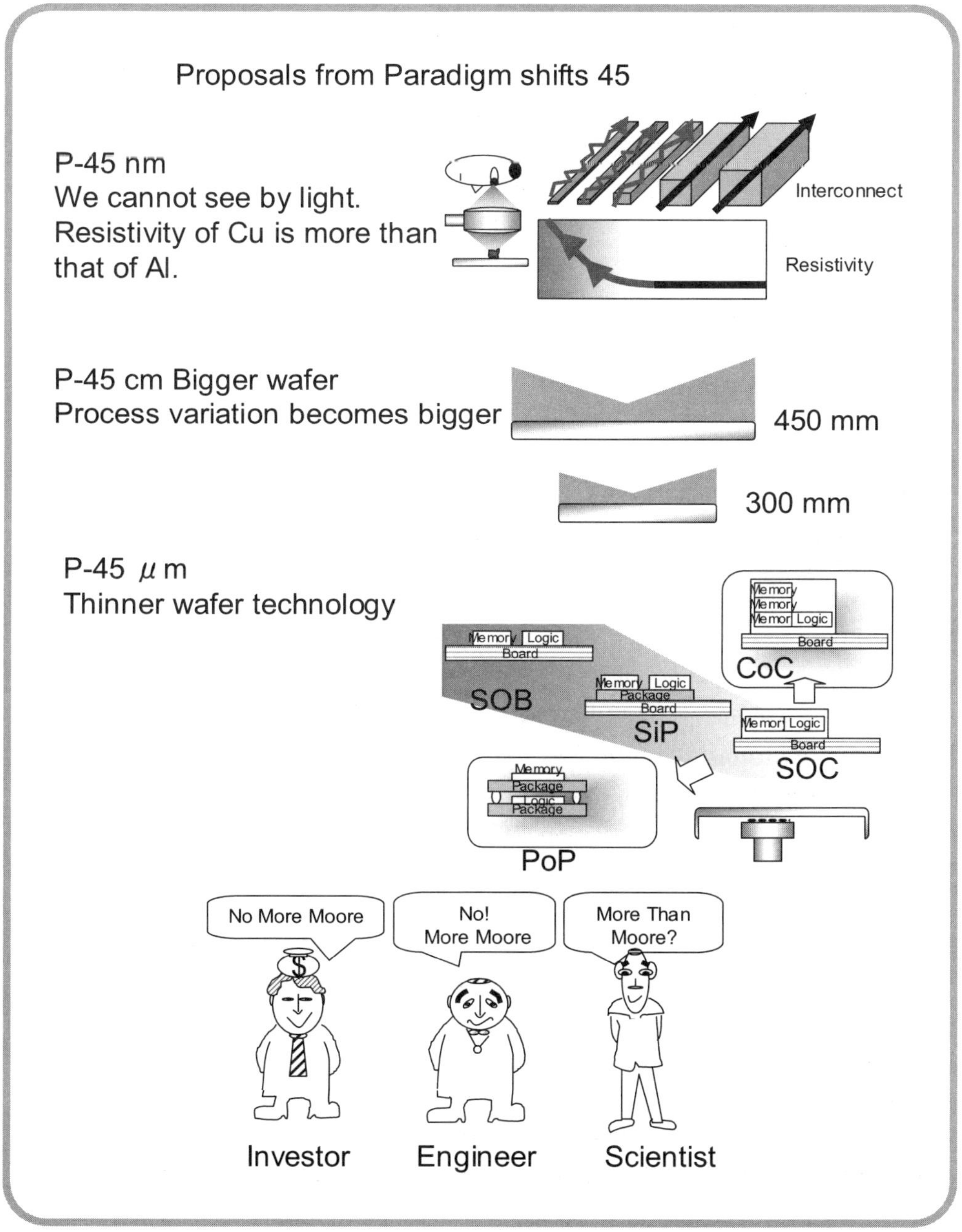

Paradigm Shift 45 followed by a paradise?

Many of the world's top investors raised questions about the issues raised in Paradigm Shift 45, and the world's top engineers and scientists found the answers to them. The results of top-level discussions [17)] are presented below.

Concerns from the viewpoint of investors

Investors are concerned about the future of scaling and wafer size transitions in terms of three points: (1) Does the investment cost increase and weigh on business, (2) Can the demand be shared by all manufacturers (unpopular products do not sell even if their prices are low), and (3) Is the timing of investments considered from the perspective of both technology and business? They are willing to invest in the semiconductor industry if the answers to these questions can be well explained to them. The answers are as follows.

Answers in light of scaling

For devices preceding the 45 nm generation, scaling has provided (1) faster processing, (2) lower power consumption, and (3) cost reduction through higher integration (Dennard's scaling theory). Thus, scaling has contributed to the prosperity of the semiconductor industry over the past 50 years (Moore's Law). For devices of the 45 nm generation and later, as shown for P-45 nm, scaling causes (1) slower processing, (2) higher power consumption, and (3) cost rise through higher integration! However, scaling to 10 nm or less is possible, and there are many ideas of reducing power consumption by using new materials/designs. Development incurs costs, but it need not cover all aspects. It is important to incorporate "necessary technologies into necessary devices." Scaling is still promising (optimistic opinion).

Answers in light of wafer size transitions

The proverb says that, if a man keeps cherishing his old knowledge, so as continually to be acquiring new, he may be a teacher of others. The introduction of 450 mm wafers can be discussed in the same way as for the introduction of 300 mm wafers in the later 1990s. Advocates for the wafer size transition to 450 mm point out that (1) wafer demand is still growing and (2) the number of 300 mm wafer plants is about to exceed 300. People reluctant (not opposed) to promote the transition have (1) a simple question about the growth of wafer demand and (2) a concern about the huge development cost of 450 mm wafer equipment, which is higher than that of 300 mm wafer equipment.

Answers in light of 3D integration and emerging research devices

In the past, a wafer size transition was driven when scaling was not achieved as planned; during such a time, efforts on scaling were continued. Thus, scaling and wafer size transitions interacted well with each other. In the current situation where both scaling and wafer size transitions have lagged, 3D integration may be employed as a solution. It offers many advantages, such as reducing the wafer size and solving the RC delay problem in multilayer interconnects (especially, global interconnects). It is time to adopt 3D integration. Scientists also insist that the development of emerging research devices going beyond scaling and wafer size transitions should be emphasized (optimistic opinion).

A paradise or the end of the world?

The results of discussions over a year are summarized below. Engineers and scientists have optimistic opinions that (1) scaling proceeds as far as desired, (2) further wafer size transitions are possible if desired, and (3) the adoption of 3D integration and the development of emerging research devices should be addressed from now on. It has been found that, depending on the matching of technology and business, the semiconductor industry with the 45 nm generation and later may be taken to a paradise or an end.

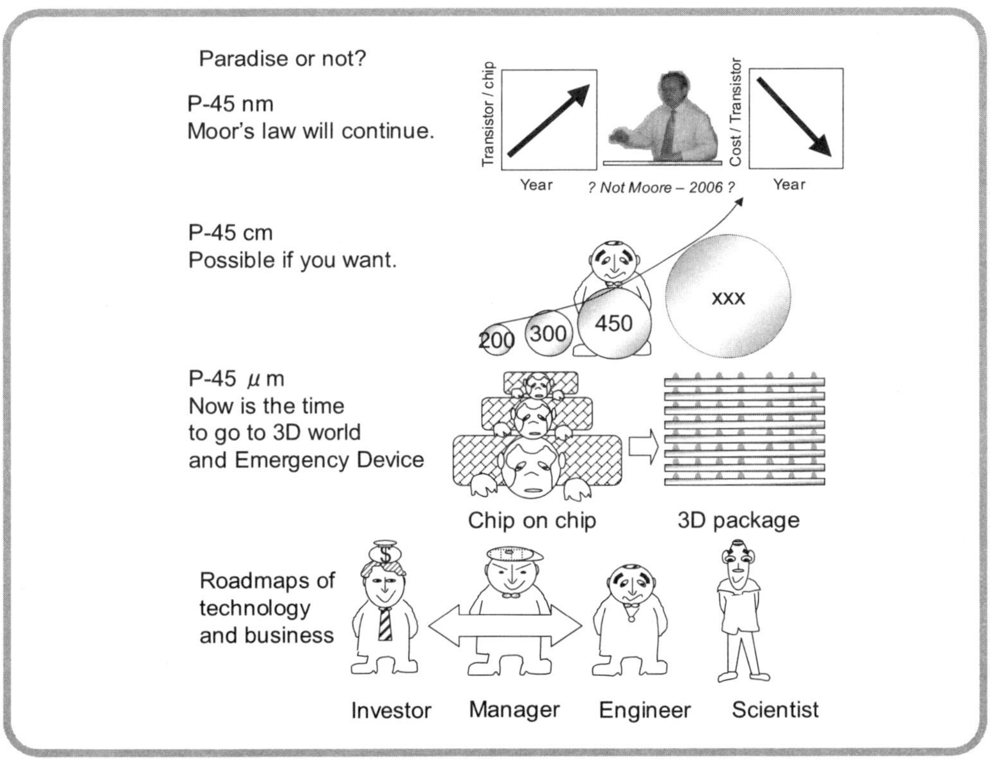

Higher and higher

High-rise condominium and multilayer interconnection (Figure 1)

It may be said that the multilayer interconnection of semiconductor devices is similar to the construction of high-rise condominiums. With scaling, the number of interconnect layers on a semiconductor device increased from two to three in the early days to eight to ten today. Why?

The performance of transistors improves with scaling as described earlier. The same principle does not apply to interconnects. If the amount of applied current is unchanged, the interconnect length must be increased to compensate for the scaled (reduced) width. To accommodate the increased length, an extra portion of the interconnect is placed on the next higher layer. In this way, the number of interconnect layers increases.

At the same time, the selection of materials changed. Although details are omitted, the interconnect material was changed from aluminum to copper for resistance reduction. Various insulating materials are also under development for lowering the dielectric constant. The material change from aluminum to copper and a lower dielectric constant have caused awful problems, including:

Copper corrosion;

Voids in copper;

Fragile and easy-to-break insulating material, and;

Voids in insulating material.

There are many voids, corroded posts, and numerous scratches…

Transistors have unsafe structures; safe multilayer interconnection like a safe condominium is demanded.

Summary of problems (Figure 2)

Problems pointed out at academic meetings are picked up for each process.

- Deposition/cure of low-k material: This material is weak and cured to sufficient strength, but it shrinks after curing!
- Etching of deposited low-k film: The drilling process generates many particles!
- Barrier deposition: Barrier layers for secure protection contain many defects.
- Plating/polishing: Plating generates voids, and polishing causes scratching and corrosion.

While there are also many other problems, semiconductor device technology is ever advancing. I wonder how far the technology will progress.

Fig.1 Similarity between interconnect and high-rise condominiums

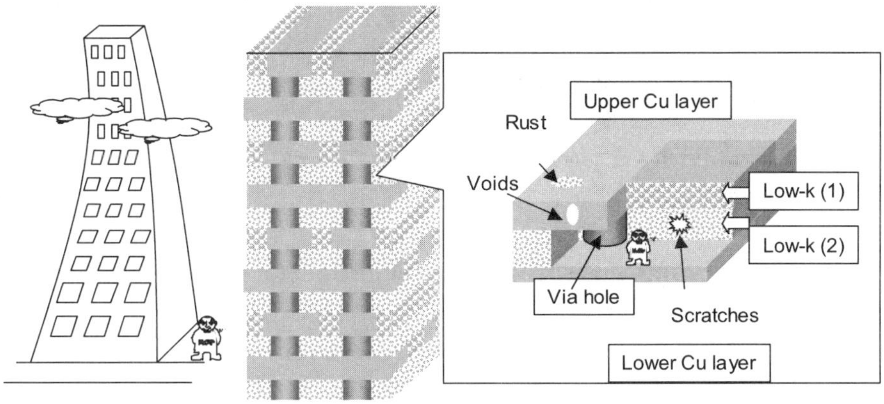

high-rise condominiums

Interconnect

The contents?

Fig.2 Various problems of interconnect

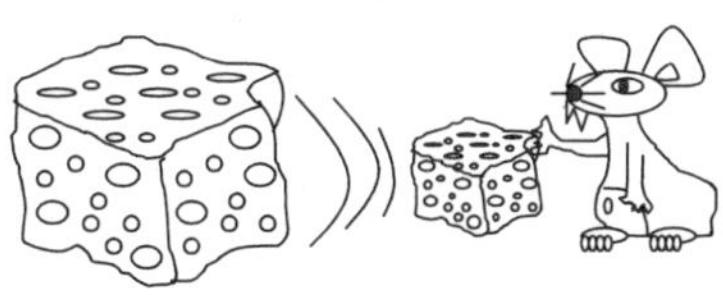

Low-k seems like cheese

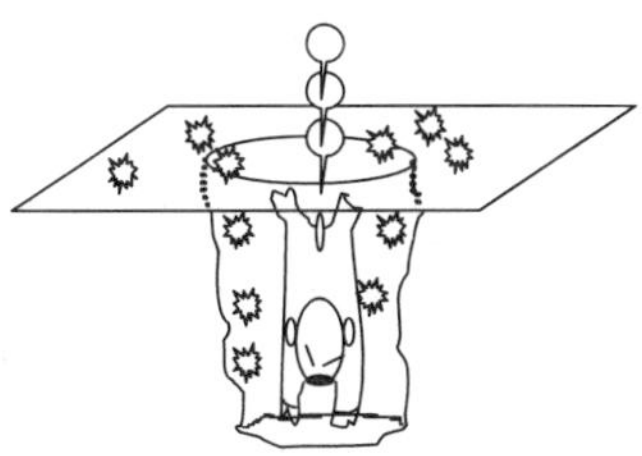

RIE is dirty

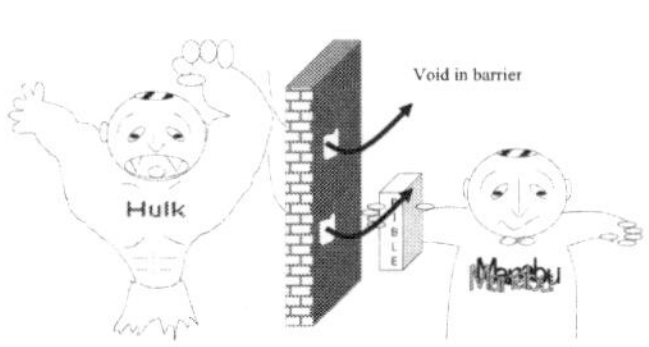

Voids in barrier layer

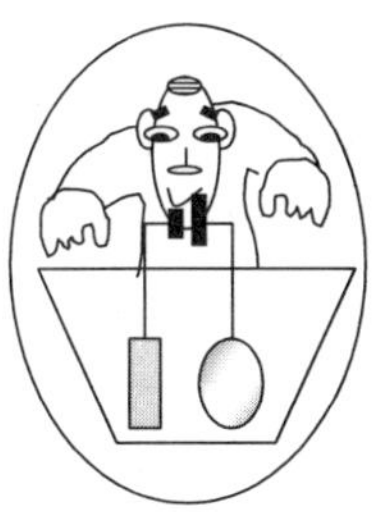

How about plating?

How about CMP?

To the end of the world

Vigorous debate at ADMETA (Figure 1)

No worry is needed regarding many problems mentioned earlier. Engineers around the world do not overlook the problems. Unlike the case of high-rise condominiums built with faked earthquake-resistance data, the engineers never commit fraudulent acts violating established standards. They are addressing the problems straightforwardly. At ADMETA (Advanced Metallization Conference), chaired by the author in the past, young engineers concentrate on solving the problems.

Application of plating/polishing to semiconductor processing (Figure 2)

Polishing has been used since the Polished Stone Age; plating dates back to ancient times. Now, they are used in highest-tech semiconductor processes.

Semiconductor devices are broadly applied to PCs, cell phones, digital appliances, vehicles, and robots. As they apply to a wider range of areas, the requirement for higher performance must be met. The improvement of device performance widens the application range, which in turn leads to performance improvement. In this way, a positive cycle has been established.

The same is true for plating/polishing. The world's oldest technologies of plating and polishing were accepted into highest-tech semiconductor processing with the most desperate efforts of plating/polishing engineers. Plating and polishing have been recognized as semiconductor processes, and many patents related to them have been created. They are a good example embodying the concept of Wet Revolution!

Looking forward to revolutionary technologies (Figure 3)

As explained earlier, multilayer interconnection is similar to the construction of high-rise condominiums. There are many challenges to be addressed, such as reinforcement, degassing, and heat release. It means that great opportunities are available for young engineers to develop revolutionary technologies.

Scaling also advances. Efforts should be made until the end of the world.

Semiconductor processes make progress with many problems.
"Necessity is the mother of invention, and desire is the father of revolutionary technologies." The first step is to identify problems.
Young engineers in this industry are expected to accomplish these goals.

Fig.1 At the academy

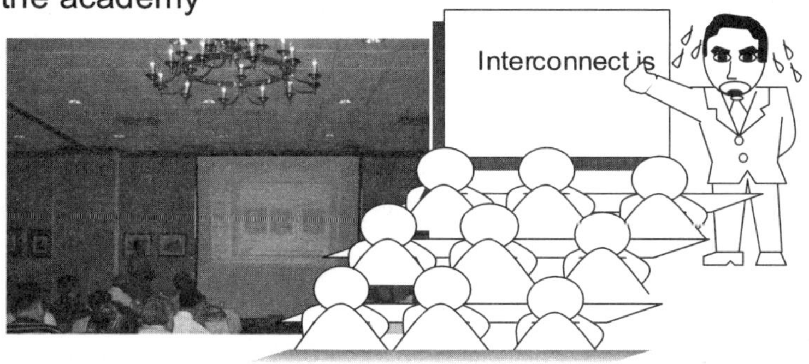

Fig.2 History of plating and polishing

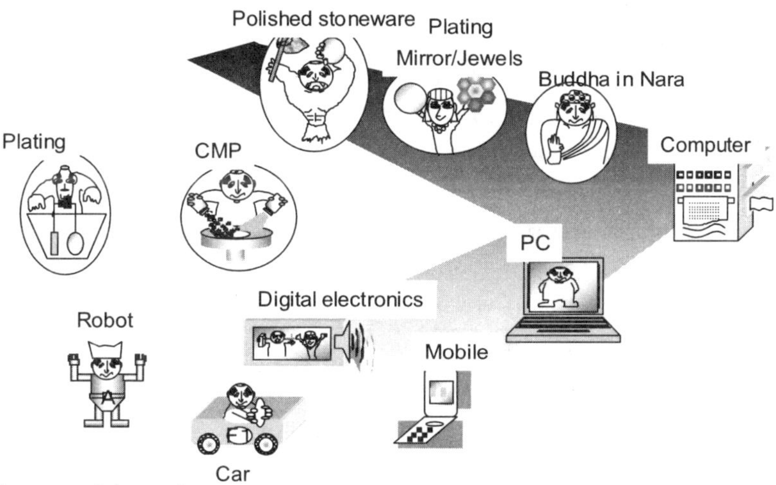

Fig.3 The world end

Emergence of the God with Divine Hands

More and more planar

Planarization technology for semiconductor devices has evolved with multilayer interconnection. Chemical Mechanical Polisher (CMP) emerged in the 1980s and was adopted because of its good planarization performance. It was first used for inter layer dielectrics and then also for metal filling (tungsten, copper, etc.). At present, CMP is essential for semiconductor processing.

Do you know the planarization level to be achieved in the future?

According to the International Technology Roadmap for Semiconductors (ITRS), an unevenness of *8 nm* is required for the 32 nm generation scheduled for 2010. *8 nm*!

Object to be planarized (Figure 1)

See Figure 2, which shows an example of Cu interconnect formation. After plating, the plated surface is non-planar. Generally, the polishing of the plated surface generates concaves, including dishing and erosion, as shown in the figure. These concaves make unevenness and must be shallowed to 8 nm or less.

Emergence of the God with Divine Hands (Figure 2)

The God with Divine Hands has emerged to solve this issue. See Figure 2.

A miracle happens when the God with Divine Hands touches the non-planar surface!

The surface is planarized at once.

That's incredible, isn't it?

Fig.1 Objects to be planarized

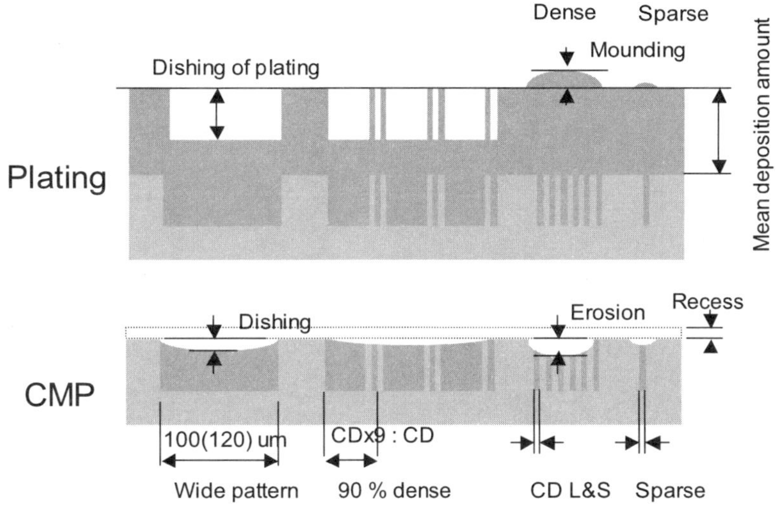

Fig.2 Surface is planarized at once

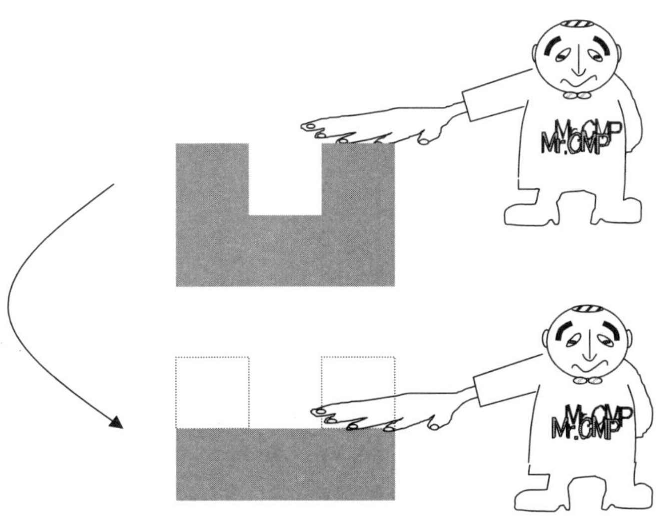

God with Divine Hands: Governing planarization technologies

CMP, ECMP, ECP, CE: A variety of planarization technologies (Figure 1)

In the 1990s, when the importance of planarization was recognized, various planarization technologies emerged. CMP emerged first, followed by Electro-Chemical Polisher (ECP) and Electro-Chemical Mechanical Polisher (ECMP). The situation became rather complicated. Furthermore, Chemical Etching (CE) also emerged, and the planarization industry entered the era of diversification.

I published a paper intended to explain the general mechanism of these technologies. The paper entitled *General Principle of Planarization Governing CMP, ECP, ECMP & CE* has aimed to help users (and students) understand the general mechanism of various planarization technologies, not focusing on theoretical development.

Figure 1 shows the mechanism of planarization technology. For more details, refer to papers listed at the end of this book. A noteworthy point is that this figure is (maybe) *beneficial* for understanding the general mechanism of CMP, ECP, ECMP, CE, and all other technologies based on the combination of them.

Divine hands: ECP using ultrapure water (Figure 2)

A planarization technology that matches the concept of *General Principle of Planarization Governing CMP, ECP, ECMP & CE* has come out. The technology, known as a process like divine hands, is based on ECP using ultrapure water [18]. It has an extremely simple mechanism with:

Ultrapure water, which offers insulation and blocks the flow of electricity, and;

Ion exchange membrane, which allows the flow of electricity.

The ion exchange membrane is immersed in ultrapure water and then is contacted with the non-planar copper surface of devices. Only portions in contact with the ion exchange membrane are dissolved. The ion exchange membrane is the essence of divine hands. The ion exchange membrane formed as a rigid (non-flexible) film contacts and dissolves only convexes on the non-planar surface, as shown in the figure. Thus, planarization is achieved.

What a wonderful process it is!

Fig.1 General principle of planarization

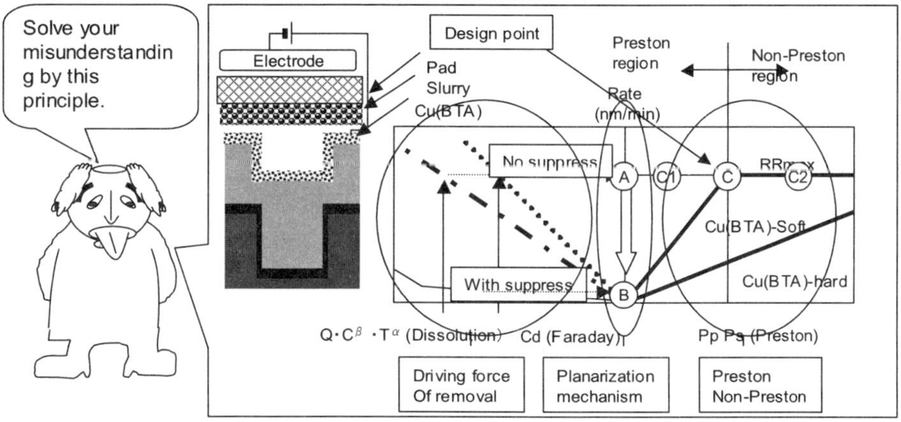

Fig.2 Ultra Pure Water ECP = God with Divine Hands

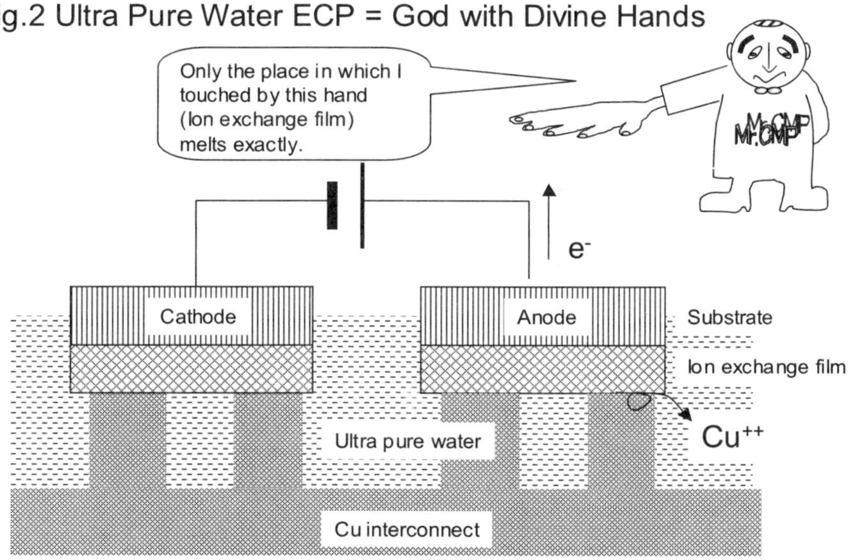

Superfast God: Faster rotation

Faster and faster (Figure 1)

CMP aims to reduce the initial step height on the semiconductor device surface as described earlier. Experience has shown that a higher polishing rate (faster wafer/table rotation) improves planarization performance. For example, Figure 1 shows that the initial step height is 100 nm. When the wafer is polished at a table rotation speed of 30 rpm, the final step height is 30 nm. When the wafer is polished at 60 rpm, the final step height is 15 nm.

It may be natural to think that faster rotation improves the planarization.

I had the same thought and believed that this idea was Something New.

Turbo molecular pump running at 50,000 rpm (Figure 2)

Ebara has expertise in turbo molecular pumps. Turbo molecular pumps are vacuum pumps for producing ultrahigh vacuum. Their blades are supported by a magnetic bearing and subject to little friction, allowing pump operation at 50,000 rpm.

However, there is a problem.

A turbo molecular pump operates under vacuum and with minimum load, while polishing involves loading machinery. Problems may occur if the pump runs at 50,000 rpm under loaded conditions.

In fact, this concern has been resolved.

Ebara is also a manufacturer of precision vibration isolation tables based on magnetic bearings and has expertise in load handling.

The combination of rotating machinery (turbo molecular pump) and loading machinery (precision vibration isolation table) may provide a rotation speed of about 1,000 rpm, not to say 50,000 rpm!

The *ultrahigh speed CMP project* was launched with the participation of the designers of the turbo molecular pump and precision vibration isolation table, but some anxiety remained...

Fig.1 Faster and faster

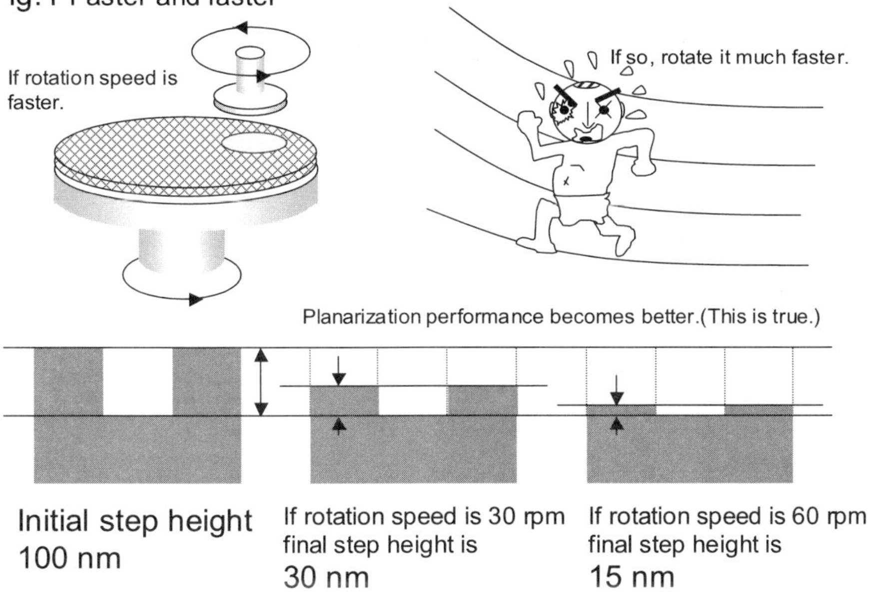

Fig.2 Ultra-high speed CMP project started.

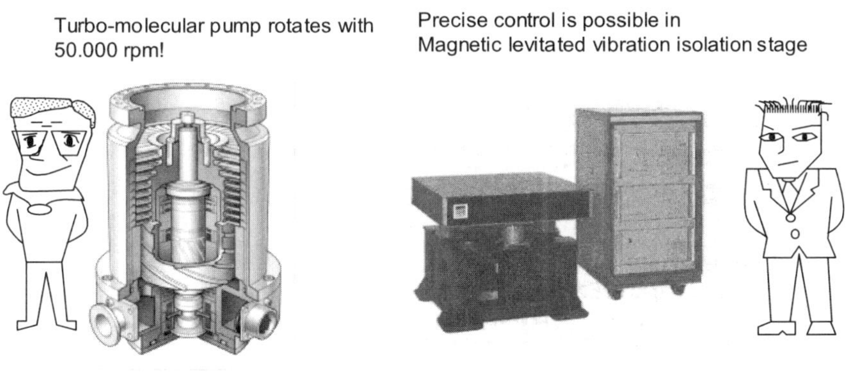

Superfast God: Too fast...

Realization of superfast CMP (Figure 1)

See Figure 1.

After overcoming various difficulties, a CMP system was created that had a polishing carrier (wafer holder) with a mechanism equivalent to the turbo molecular pump and could run at least 1,000 rpm [19)].

The carrier is pressed against the pad on the table. Therefore, it may be forced to be dragged or lean forward. To prevent these problems, the carrier position is controlled with four sensors in the radial direction (at angular intervals of 90 °) and two sensors in the axial direction.

Furthermore, possible carrier vibrations due to high load are eliminated with the precision vibration isolation table.

The designers of the turbo molecular pump and precision vibration isolation table had no experience in dealing with CMP applications, but safe operation at 1,000 rpm was verified.

Then, the project proceeded to the stage of testing under process conditions.

It was expected that the step height might be zero (I could not help but laugh).

Nothing happened?? (Figure 2)

Oh, why? The step height on the semiconductor device surface remained entirely unchanged.

No planarization occurred.

At first, we suspected the possibility of measurement errors.

Again and again, the test was repeated...

However, the result was the same.

No planarization occurred at a rotation speed of 1,000 rpm!?

Subsequently, the reason for this result was revealed.

It was found that the carrier hydroplaned (floated on a layer of liquid) due to the excessive rotation speed.

This phenomenon can be solved.

The solution is not just to lower the rotation speed.

Indeed, a rotation speed of 1000 rpm is too high, and the problem is solved by determining the appropriate rotation speed and by analyzing the liquid layer.

The *ultrahigh speed CMP project* continues in the future. The Superfast God does not descend easily.

Fig.1 So, we tried to rotate CMP 1000 rpm.

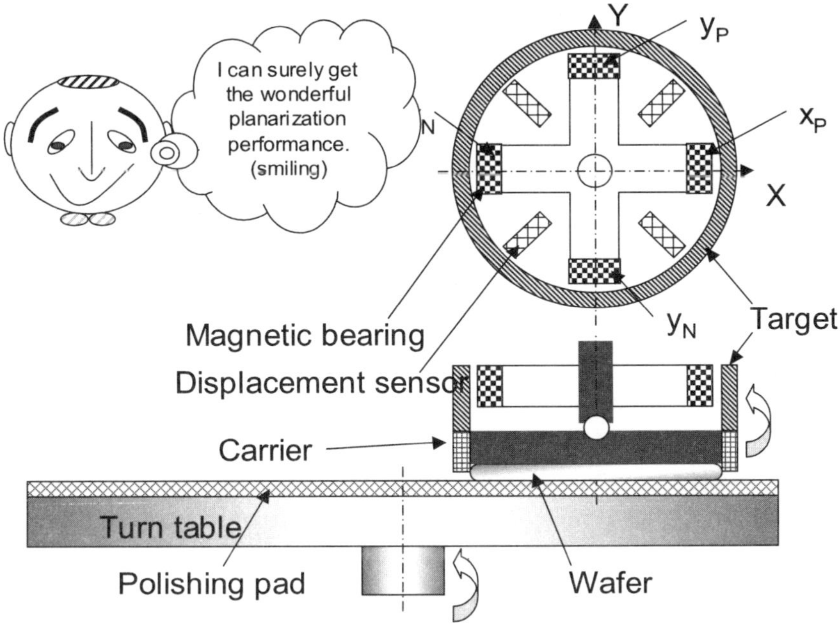

Fig.2 Anything regrettably did not happen.

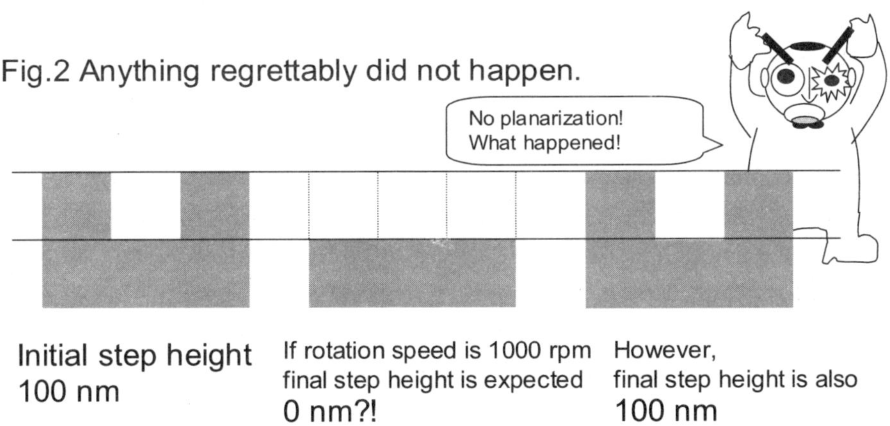

God with Metabolic Syndrome: Exceeding physical limits

Consequence of scaling (Figure 1)

The scaling of semiconductor devices reduces the line width, making barrier layers and seed layers for copper plating narrower. For the 45 nm generation, the thickness of barrier and seed layers is 3.3 nm.

Barrier (typically, made of Ta) and seed (copper) layers are deposited using sputtering systems. As shown in Figure 1, their thickness is not conformal. It is highly possible that interconnect copper lodges at the entrance and is insufficiently deposited/plated, forming voids. Even if gap filling deposition is possible, the space for Cu interconnects reduces with barrier deposition. What measures should be taken?

Thin and conformal (Figure 2)

The technology of depositing thin and conformal barrier and seed layers has developed. Traditional chemical vapor deposition (CVD) or atomic layer deposition (ALD) allows conformal deposition. These deposition methods still have room for improvement, and there are more significant and fundamental problems to be addressed.

Terminal effect (Figure 3)

The problems are electron scattering and the consequent terminal effect.

See the right graph in Figure 3. As the Cu line width decreases below about 50 nm, the electrical resistivity of copper significantly increases. This phenomenon is known as electron scattering in copper. Copper has replaced aluminum because of its lower electrical resistivity, but the electron scattering negates the advantage of copper. In fact, a physical limit is encountered.

There is no doubt that this problem affects not only interconnects with a line width of 50 nm but also seed layers with a thickness of 3.3 nm. A seed layer may have an extremely high resistivity. Electrodes for copper electroplating are typically made at the wafer edge; if the electrical resistivity is different between the wafer center and edge, the gap in the current level leads to differences in the deposition amount. The generation of differences in the deposition thickness between the wafer center and edge is called terminal effect. It poses a serious problem in electroplating deposition.

Fig.1 Seed thickness in accordance with scaling

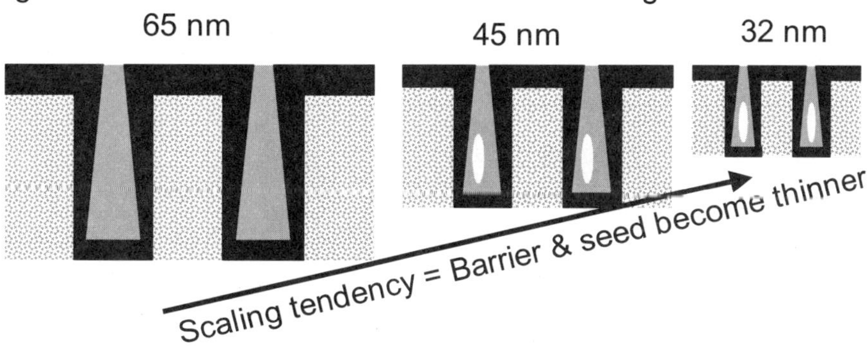

Fig.2 Barrier and seed layers become thin and conformal.

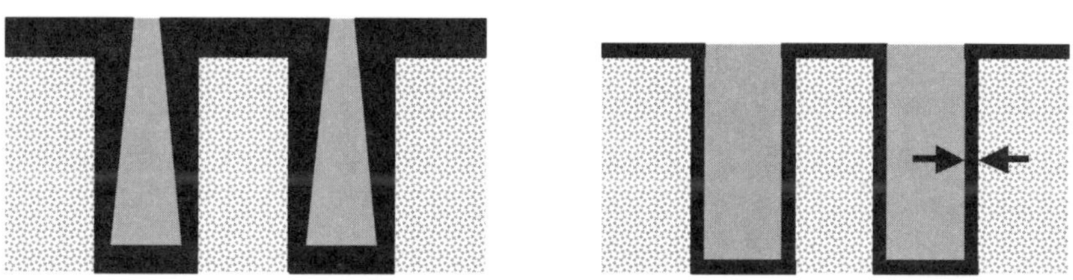

Fig.3 Electron scattering effect of interconnect and seed layers

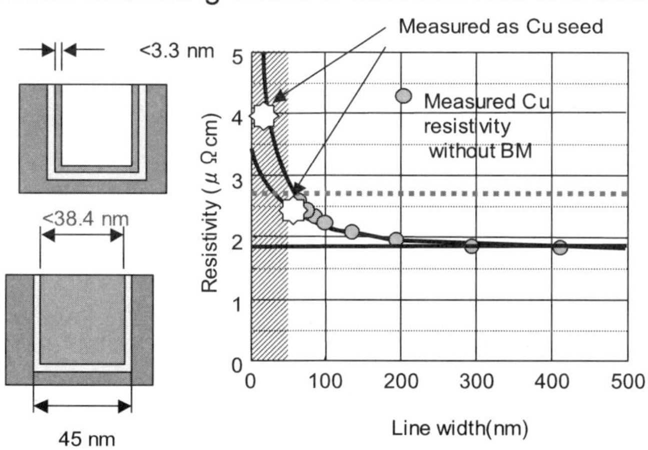

God with Metabolic Syndrome: Adding up with a dummy

Measuring body weight everyday (Figure 1)

Recently, my weight started going up again.

I have been scolded by Kumiko, my beloved wife, and may be placed on a restricted diet. So, I have decided to measure my weight everyday.

Last month, my weight was 70 kg. This month, my weight is…

Really!? 77 kg? My weight has increased 10 %!

An excuse is required. My cousin is a giant weighing over 100 kg and concerned about metabolic syndrome. The sum of his and my weights is;

(70+100) kg for the last month, and;

(77+100) kg for this month. The increase over the last month is calculated by (177) ÷ (170) x 100 = 4.1. By rounding down the value, the increase is only 4 %. Is an increase of only 4 % acceptable?

"No way!" said Kumiko.

Development of dummy plating (Figure 2)

My cousin's weight (100 kg) corresponds to a dummy material used for dummy plating [20].

Suppose that the seed resistance is 5 Ω at the wafer edge and 10 Ω at the wafer center. This difference of 5 Ω represents the terminal effect caused by the thin seed layer.

At the wafer edge in which the resistance is lower than that at the wafer center, a larger current flows, forming a thicker deposition of copper in accordance with Faraday's law. The deposition at the wafer edge is twice as thick as that at the wafer center, producing a difference of 100 % in the thickness.

This problem is addressed by inserting a *metabolic dummy*. If the dummy material has a resistance of 100 Ω, the total resistance is (100 + 5 = 105 Ω) at the wafer edge and (100 + 10 = 110 Ω) at the wafer center.

The difference is (110 ÷ 105 x 100 = 4.7 %).

Wonderful! The difference decreases from 100 % to only 4.7 %.

Dummy plating is performed in this manner.

The dummy plating process is a gift from the God with Metabolic Syndrome.

Fig.1 How to use of the dummy.

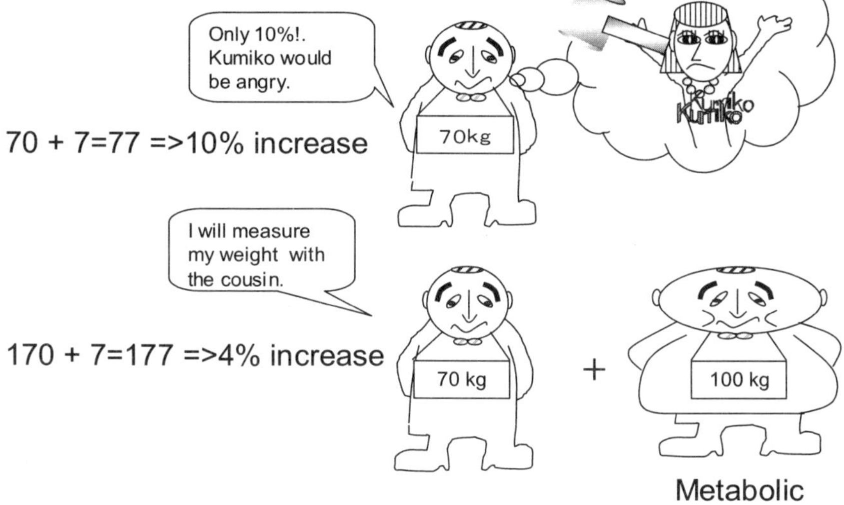

Fig.2 Dummy plating

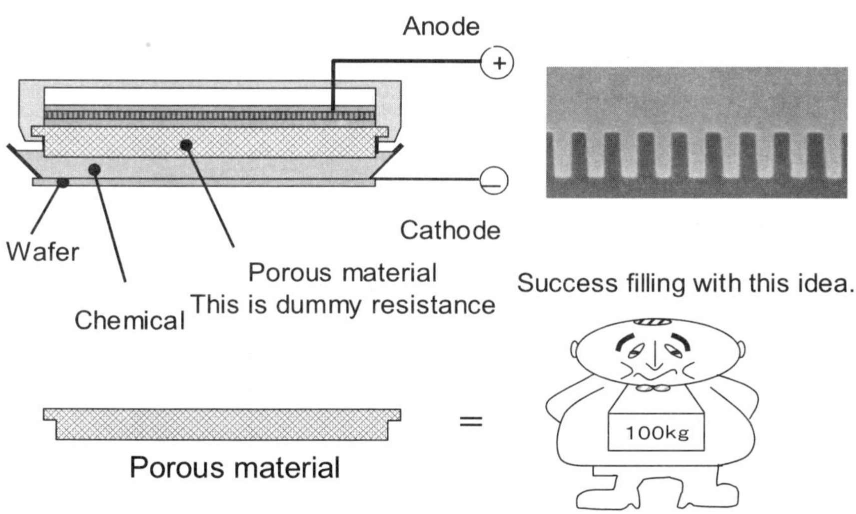

Porcupine God: Cutting waste

Inlay technique for art craft in Kyoto (Figure 1)

Do you know the inlay technique? Kumiko has a favorite jewelry box decorated with inlay work. See Figure 1. First, the surface is engraved with a desired pattern. Next, (plenty of) gold powder is poured over the surface and rubbed into the engraved surface carefully. Then, what is obtained?

It is amazing (but not so much) that the pattern appears in gold.

However, such use of gold powder is wasteful, isn't it?

Most of the extra gold powder is reused, but some may be wasted. A technique that fills only patterned grooves with a material is more useful.

Such a technique is provided by the *Porcupine God.*

As you may know, the inlay technique is equivalent to the damascene process for Cu interconnection on semiconductor devices.

Minimizing the cost of slurry (Figure 2)

A disadvantage of the damascene process is the cost of slurry used for CMP. Every effort has been made to reduce the use of slurry since its cost is high. A CMP process using a smaller amount of slurry is much welcomed. This matter has also been addressed in terms of plating deposition.

See Figure 2.

The conventional plating process requires the amount of plating shown in the figure, depending on the device pattern. After plating, copper is polished to planarize the non-planar surface. The larger amount of polishing involves the higher cost of slurry. Thus, efforts have been made to reduce the amount of plating.

There was an idea that a planar surface could be obtained with a smaller amount of plating by performing polishing and plating at the same time. At present, this idea is dismissed since polishing during deposition results in the entry of particles into deposited film.

A different approach is needed.

This need is met by the *Porcupine God.*

Fig.1 Inlaying (Damascene) technique

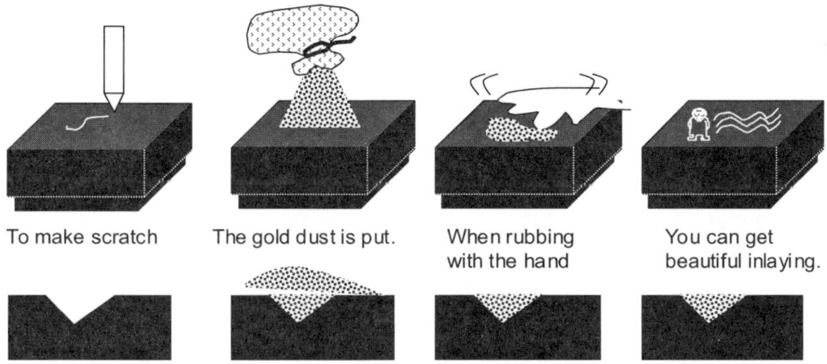

Fig.2 To decrease expensive slurry cost

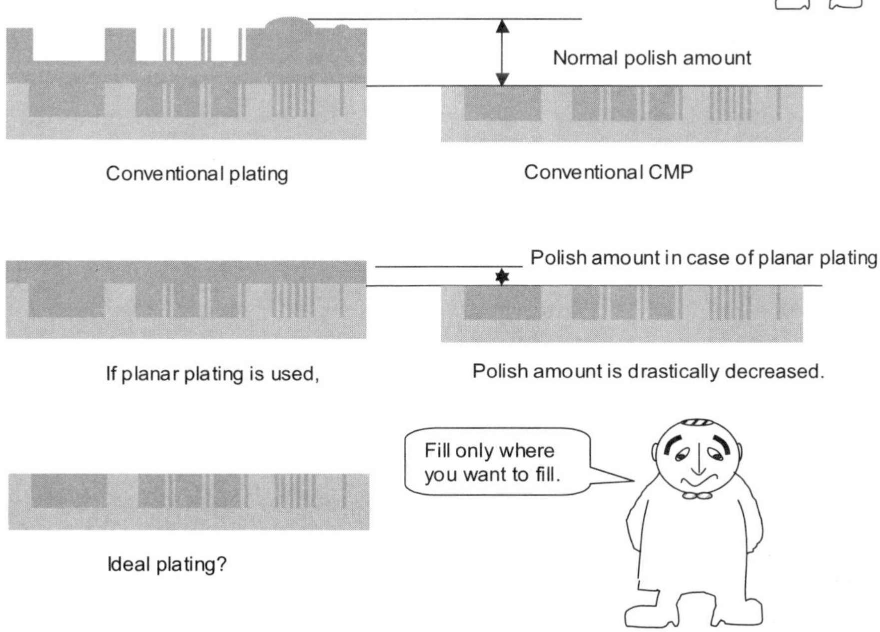

Porcupine God: Infilling

Porcupine process (Figure 1)

The *porcupine process* is described below.

It is explained with reference to Figure 1.

[Plating steps]

Step 1: The process starts with a substrate having a copper seed layer. First, a porous pad is placed on the substrate.

Step 2: Plating starts. Copper is plated/deposited onto the trench and field areas and into the pores of the pad.

Step 3: Then, the trench area is completely filled with copper.

Step 4: The pad is removed. The copper plated into the pores forms Cu posts, which are like spines of a porcupine. The *porcupine process* is named after this shape. See the cross-sectional photograph at bottom left.

[Etching steps]

Steps 5 and 6: The posts are removed through etching. Etching is effected homogeneously; with the amount of etching equal to the radius of each post, the posts are eliminated, and the field area is recessed to the depth equivalent to the post radius. Attention must be paid not to etch copper in the trench area.

Step 7: It is desirable that the etching process ends with a small amount of copper left on the field area. See the cross-sectional photograph at bottom right.

[Polishing steps]

The remaining small amount of copper is removed through CMP.

Let's examine the result. Compared to the conventional plating deposition method, the amount of polishing is about 1/10. Accordingly, the use of slurry can be reduced to about 1/10, achieving a substantial cost reduction.

Since the surface is originally planar before polishing, the performance of planarization through polishing is much higher. The porcupine process provides great improvements in both performance and cost efficiency.

These improvements are achieved thanks to the Porcupine God, but there is still a problem.

In Step 4, the pad cannot be removed smoothly, as if a jacket put over porcupine spines could not be worn again. I wish that this problem be solved…

Fig. 1 Porcupine process

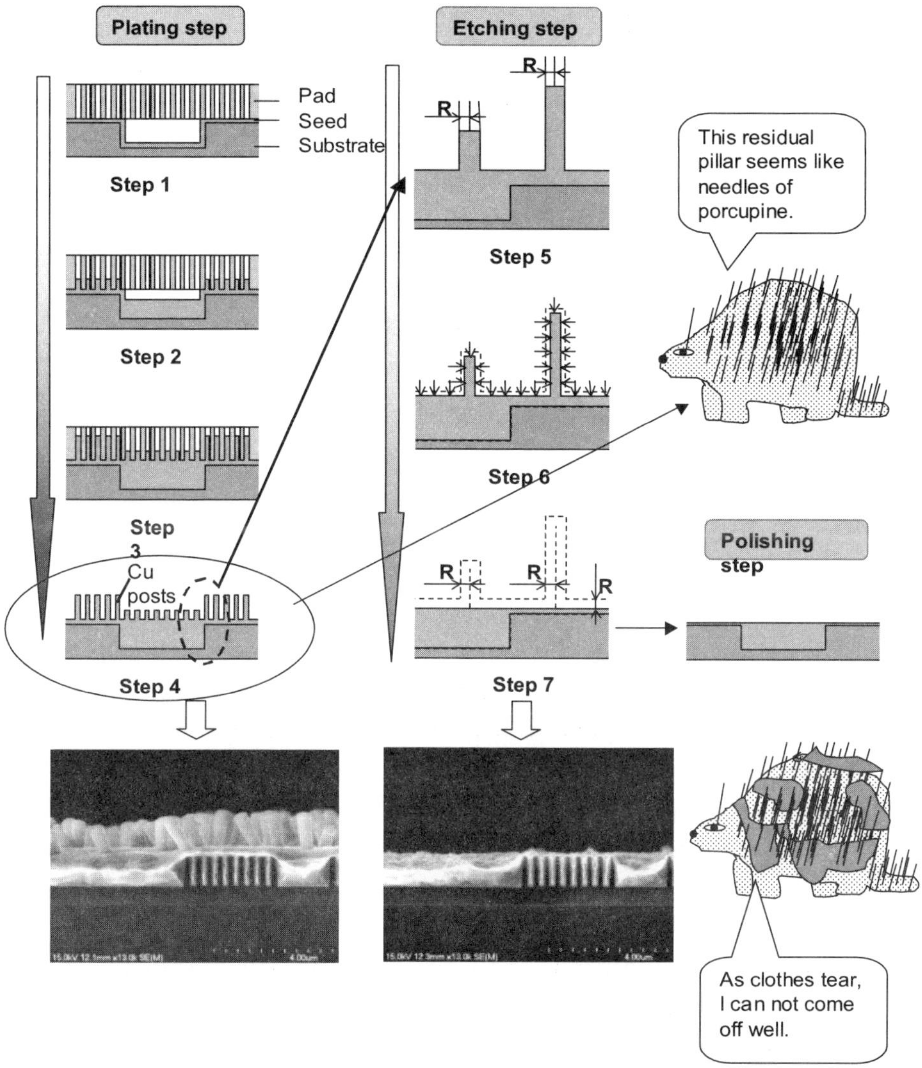

Violent God: Sign of descent

Basics of cleaning (Figure 1)

"Oh, Manabu, thank you for cleaning the floors, but would you use a clean cloth for wiping?" I was scolded by Kumiko.

Our house is not large, and I cleaned the first and second floors using the same cloth without washing it. Indeed, the floors I wiped were not clean…

This episode teaches the basics of cleaning (how dare I say such a thing?). Floors cannot be cleaned using a dirty cloth. Instead, floors get dirtier when they are wiped with a dirty cloth. I dare to say that the same principle is applicable to semiconductor cleaning.

Sweeping an ink bottle (Figure 2)

The semiconductor device surface to be cleaned is made of multiple materials unlike the floors in my house. As shown in Figure 2, it comprises oxide film, barrier material, and interconnect material. Even with copper interconnects, the surface is polished with a single brush (roll brush cleaning), although copper contamination should be avoided. When comparing Cu interconnects to an ink bottle, it is unreasonable to sweep the ink bottle while polishing oxide film!

Semiconductor device cleaning is difficult. As a solution to this problem, traditional non-contact cleaning, including jet cleaning, is in the spotlight. However, jet cleaning is inferior in cleaning performance to brush cleaning. The development of jet cleaning technology with the advantages of non-contact cleaning and high cleaning performance has been demanded.

Cleaning with rock-piercing cavitation (Figure 3)

Aside from the semiconductor industry, there are a variety of cleaning technologies. One of them is cavitation jet. Cavitation is (notoriously) known to damage ship propellers, etc. with great pressure generated by the collapse of cavities (vacuum voids) in fluid. In some industries, the destructive force of cavitation has been applied to cleaning. Cavitation cleaning is a very *violent* technology used for removing shells from ships or piercing rocks under water.

Can cavitation be applied to semiconductor device cleaning?

This idea is a sign of the descent of the Violent God.

Fig.1 With a clean dust cloth

Fig.2 If you sweep an ink bottle?

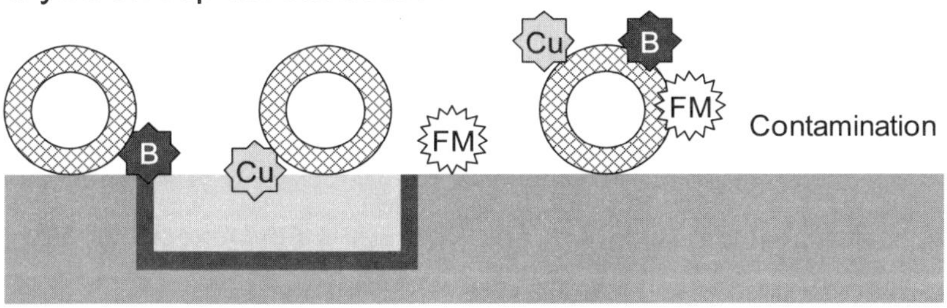

Fig.3 Non-contact cleaning

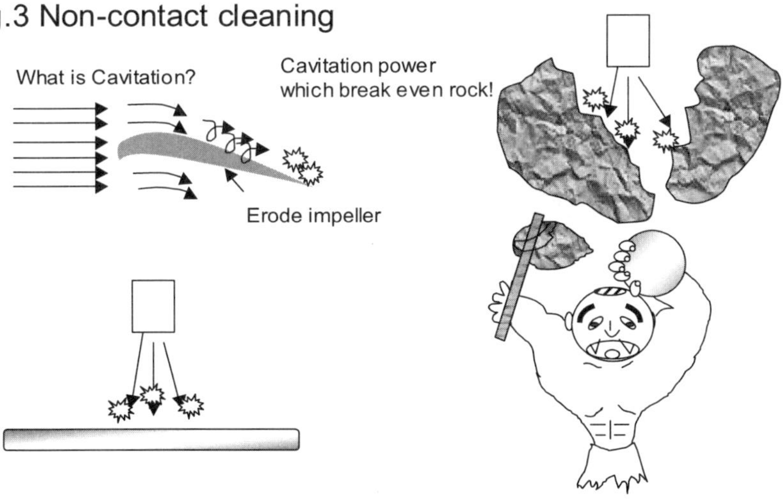

Violent God: Cavitation jet

Nozzle for cavitation formation (Figure 1)

First, it was checked if the nozzle for semiconductor device cleaning could produce cavitation [21]. In fact, Ebara has many cavitation experts, who usually deal with cavitation-related complaints e.g., that "the blades were damaged by cavitation." All of them were interested in the positive utilization of cavitation and willing to cooperate in the project. The outcome of the project is the nozzle shown in Figure 1.

The finished nozzle is based on an extremely simple mechanism.

The nozzle produces the flow of normal low-pressure water at the circumference and the jet of high-pressure water at the center.

Vortices form along the boundary between low-pressure water (slow flow) and high-pressure water (fast flow), producing vacuum (cavities) at their centers. The cavities collapse on the wafer surface and generate destructive pressure. This pressure is used with the jet flow to remove particles.

Implementation of cavitation cleaning on a CMP system (Figure 2)

The cavitation jet technology was implemented on a CMP system. For comparison with typical jet cleaning, two types of cleaning schemes were examined. First, a large amount of slurry particles used for polishing were removed by roll brushing (primary cleaning). Next, cleaning was performing by the typical jet or cavitation jet while conducting measurement with an inspection system. What results do you expect?

First of all, it is of interest whether or not cavitation damaged (!) the wafer.

Off course, not. For the cavitation jet, the cavity size and the amount of cavities generated can be regulated. Thus, the destructive force of cavitation may be adjusted to a violent level that pierces rocks like the Incredible Hulk or to a level that just removes particles on a wafer.

Our attempt was like picking up eggs with a crane.

Cleaning results (Figure 3)

Look at the results. The cavitation jet completed cleaning faster than the typical jet did. This advantage is attributable to various factors, but details are omitted here. I am happy that the results are good!

Fig.1 Cavitation jet

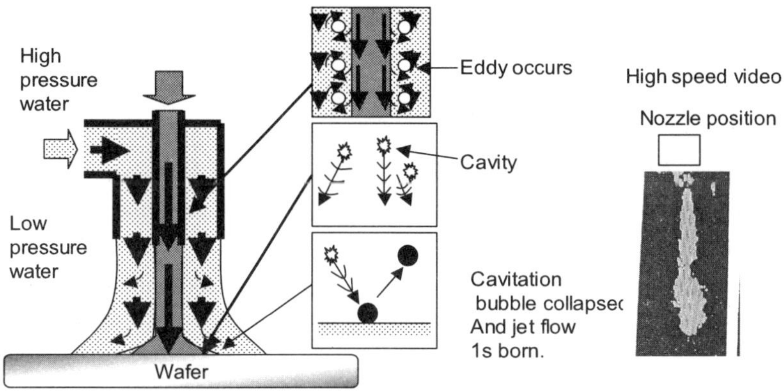

Fig.2 Cleaning tool

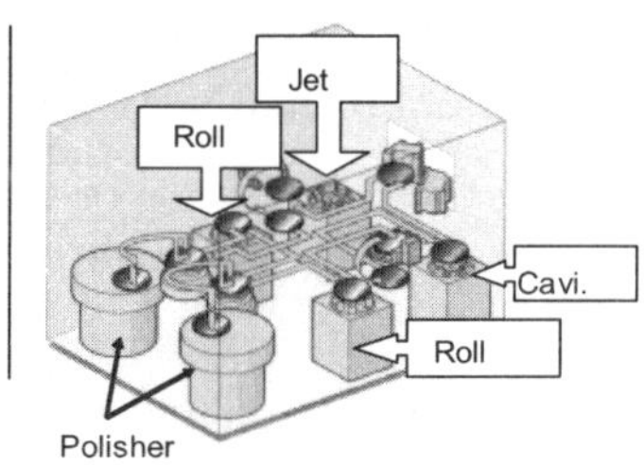

Polish -> Roll -> Normal jet ->Measure

Polish -> Roll -> Cavitation jet ->Measure

Fig.3 Cleaning results

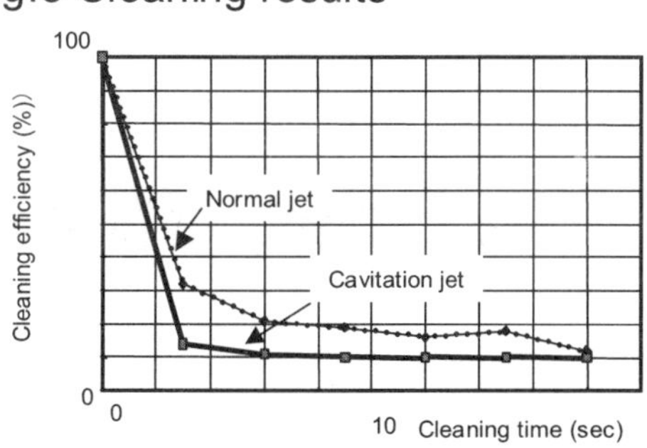

It looks like calm,but actually have a big power

Second Life and real life

I was greatly surprised when I watched a TV program last year. The TV program featured a virtual world, Second Life. I have belonged to the IT industry and known Second Life. However, I was surprised at the progress of the virtual world.
The virtual world started as a kind of game.
For example, a man in his 40's who live in a 4.5 mat room can appear as a nice woman in Second Life. In turn, a man with whom this woman talks in Second Life may be a woman in the real world. This example is somewhat terrible, but the fact described below is significant.
Famous car companies and shoe brands has opened virtual shops. Customers visit these virtual shops for test-driving or trying shoes before buying products in the real world. This virtual world is approximate to the world of *The Matrix* (movie series). However, the virtual world still includes some unnatural aspects.
Do you think that Second Life expands and approaches to the real world as IT advances?
IT evolves through synergy between the development of hardware (devices) and the demand for software. In the future, the technology may reach a level at which it is impossible to distinguish between virtual reality and reality.
An idea has hit me.
It is unrealistic to insert a plug into a socket planted on the back of the neck like *The Matrix*, but imagination may be driven by directly stimulating the brain. We have enough imagination to enjoy the Game of Life (board game). How about the fusion of IT hardware and the human brain? In this way, virtual reality that closely approximates real life may be available.

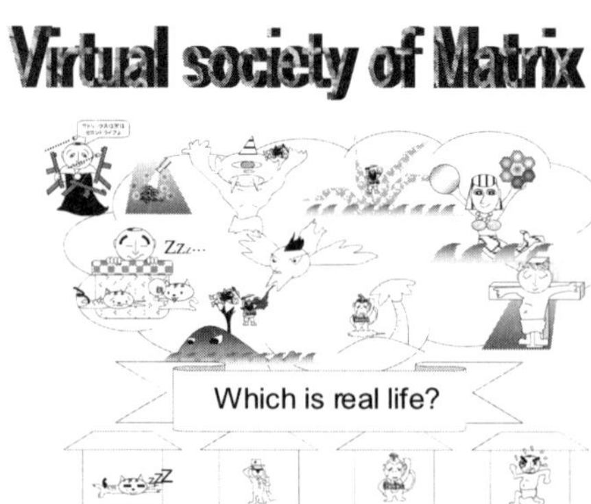

Chapter VII

Let's Look at the Forest of Technology Development!

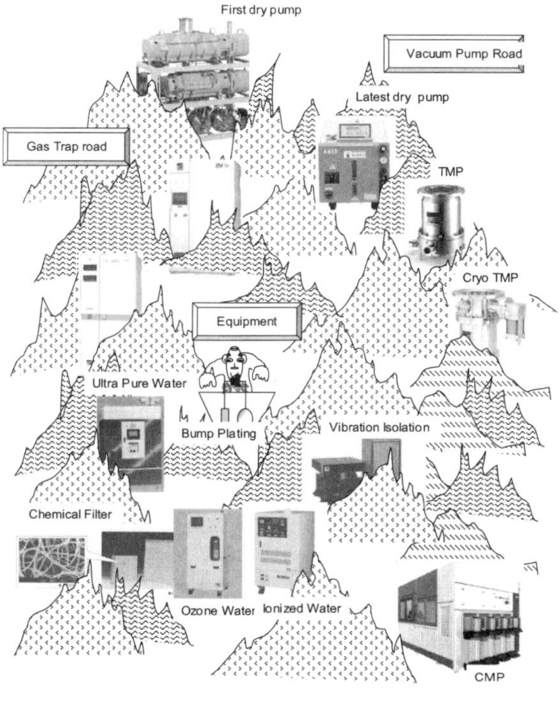

"One good turn deserves another" in development

Starting with pumps

I have been engaged in the development of semiconductor equipment (including components) for over a quarter-century. I started with vacuum pumps and then moved on to waste gas treatment systems for treating exhaust gas from vacuum pumps. Many paths branched from there.

Turning to the left, there was a path to turbo molecular pumps. Our basic technology for magnetic bearings was applied to the development of magnetic bearing turbo molecular pumps.

Going further in the path, I encountered a disadvantage of turbo molecular pumps; sufficient water discharge capacity was not achieved. Our technology for refrigerators was applied to the development of cryo turbo molecular pumps.

Path to waste gas treatment systems

Turning to the right, the path led to waste gas treatment systems.

At first, I had no idea which gases should be used and what kinds of reaction product were produced in vacuum systems for semiconductor manufacturing. It required investigations, which became valuable assets to me later. One good turn deserves another; we never know what future brings. Nothing was wasted.

We developed both dry and wet waste gas treatment systems. The path to the dry type led to PFC abatement/recovery via many twists and turns.

Great turning point to semiconductor equipment

After that, I reached a great turning point to semiconductor equipment.

Semiconductor manufacturing processes require plating. Our plating technology was applied to the development of bump plating systems. Plating is one of so-called 3K technologies: *kitanai* (dirty), *kitsui* (tough), and *keiken-izon* (experience-dependent). Later, I was rewarded by handling 3K plating.

There was no problem for me to handle another 3K technology, polishing, which is used for semiconductor equipment. I started the development of CMP.

This is just a part of my 25-year career. The forest of semiconductor equipment development is really complex. Development is like groping in the dark as described earlier. I groped in the dark forest. However, looking back at it later, I found "One good turn deserves another"; efforts lie behind achievements.

Overview of R&D forest

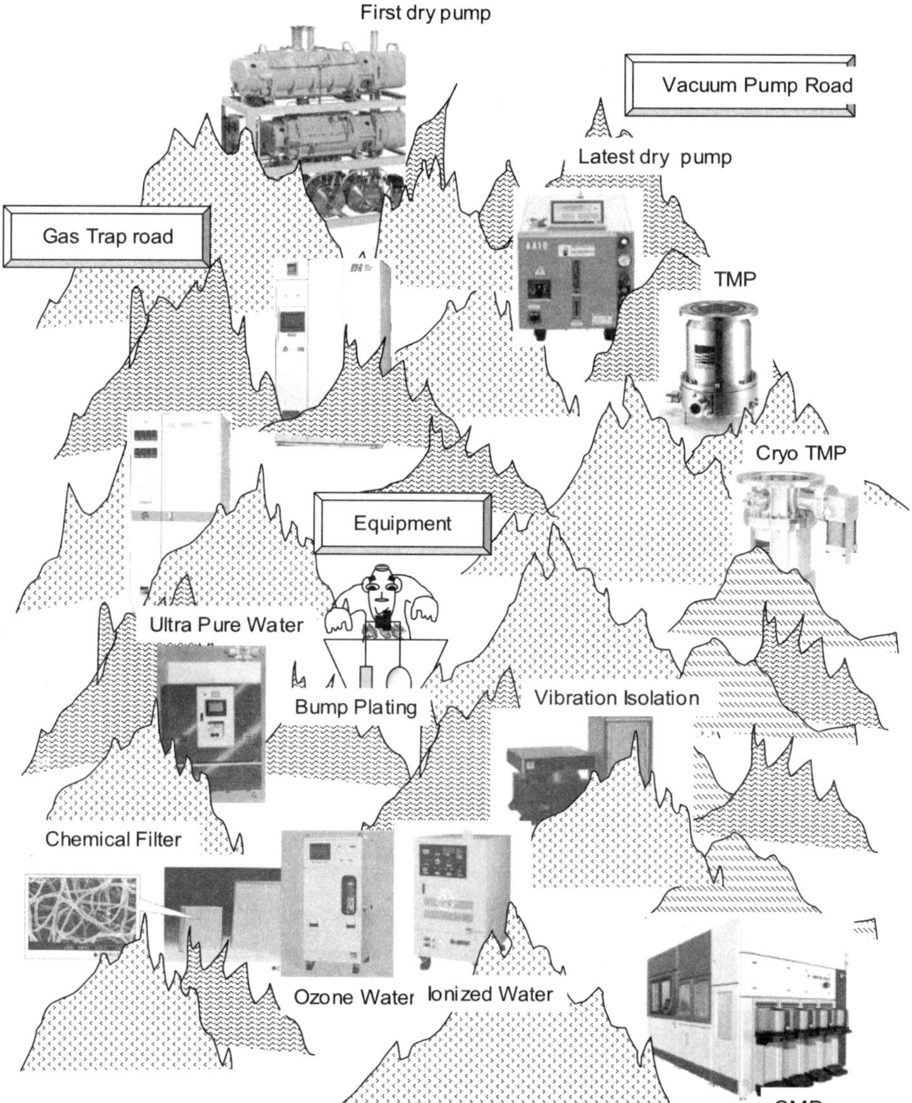

Development brings serendipity.

Like a windfall (Figure 1)

I always have to find where things are because I am disorganized. The other day, I looked everywhere in the house for my pension account book. I could not find it, but I found unexpected things, including sugar bowl savings.

I have experienced that development makes unexpected gains like a windfall. In the course of struggling for developing something, I made great discoveries instead of achieving what was expected.

Unexpectedly discovering valuable things while looking for something else is called serendipity. This is the word my master taught me.

Dummy plating as an alternative to brush plating (Figure 2)

Dummy plating was discovered by serendipity.

Our original idea was to use stamp plating like brush plating. By stamp plating, you can perform as much plating as you like only in the desired area, and no seal for a wafer is required. However, with no suitable material for stamp plating found, plating was performed by taking a small distance over a wafer. We thought that plating solution could be retained by surface tension.

The idea of using surface tension did not function well, but using a porous material instead of a brush opened the way to unexpected results.

As you may understand by reading the section for the *God with Metabolic Syndrome*, there was the problem of thinner seed layers that degraded plating uniformity. Surprisingly, dummy plating showed excellent uniformity. Why? The dummy material used instead of a brush provided dummy resistance. Since we also had the idea of using dummy resistance, ideal results were obtained.

From ultrapure water to functional water (Figure 3)

We developed an on-site ultrapure water system based on the concept of supplying the required amount of water of required quality. This system did not sell well, but we understood that the demand for functional water, such as ozone water and ionized water, was increasing. We also realized that the system supported almost all functions required for the generation of functional water. We just removed unwanted elements from the system to produce ozone and ionized water systems. Development was completed in a very short time.

Fig.1 Like a windfall

Fig.2 Born from brush

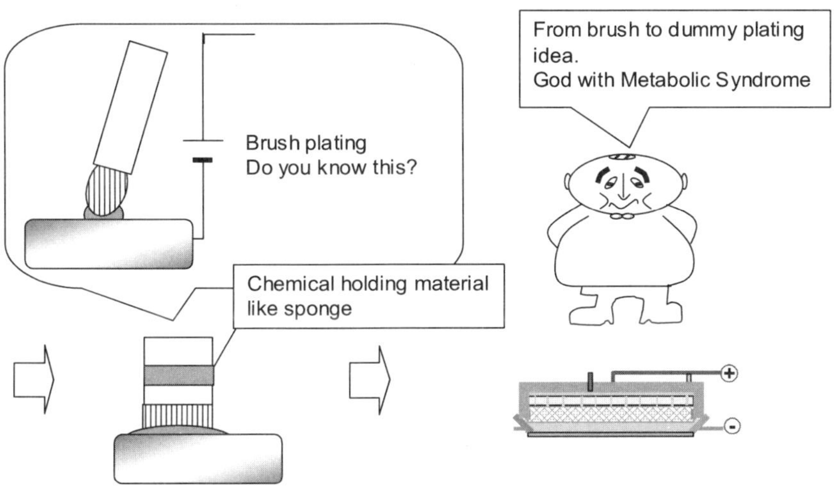

Fig.3 From ultra-pure water to function water

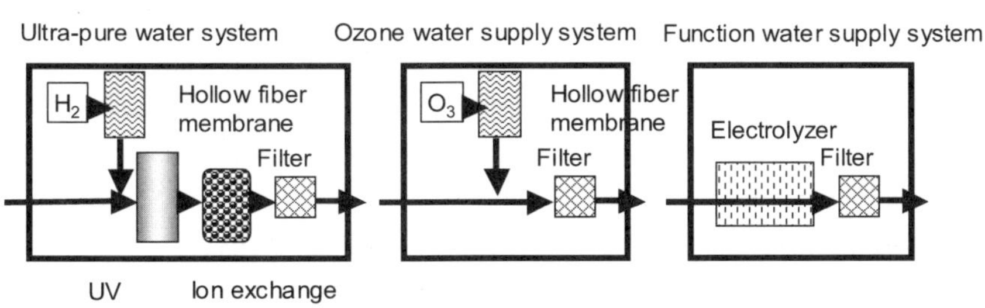

Development requires looking ahead 10 years.

My father's bonsai (Figure 1)

My father started bonsai at age 50. He had no background of growing flowers and grasses, let alone bonsai. First, he bought a camera.

???

Why camera?

He could have bought a pot or a ready grown bonsai tree...

Next, he bought a kiln for making pots. He made a pot before growing bonsai.

Eventually, it was time for growing bonsai. He took many pictures of the tree he planted. It was laughable to see him taking so many pictures of the tree as he was still an amateur.

It happened three years later, only three years later.

He published a book entitled *Bonsai ni 10-nen wa iranai: 3-nen de kanseisuru bonsai* (Only 3 years, not 10 years, are enough to grow a bonsai tree).

I was surprised to see the book containing the three-year detailed record of the bonsai tree grown in the homemade pot with pictures.

He must have had a clear vision when he started bonsai three years ago.

I was bowled over.

Development requires looking ahead 5 to 10 years (Figure 2)

Although I bossily explain what development is, I do not have a clear vision for the future as my father did. Things happen before I realize them. However, looking back the past, I convince myself that, in some cases, I did look ahead unconsciously. Such cases were taken as "One good turn deserves another" or could be rephrased as serendipity when I went to different paths.

Changing the type of vacuum pumps to the dry type caused undesired gas flow. Neither plating nor ultrapure water systems sold well. However, based on these developed technologies, we were able to develop innovative products, such as gas treatment systems, CMP, and functional water systems.

Even at my age, there are still many things I reflect on.

I always want to make people puzzled like my father, "Why does Mr. Tsujimura need this for development?" Three years later, they will find out results and think, "Oh, he must have expected these results."

Fig.1 Planning from the beginning

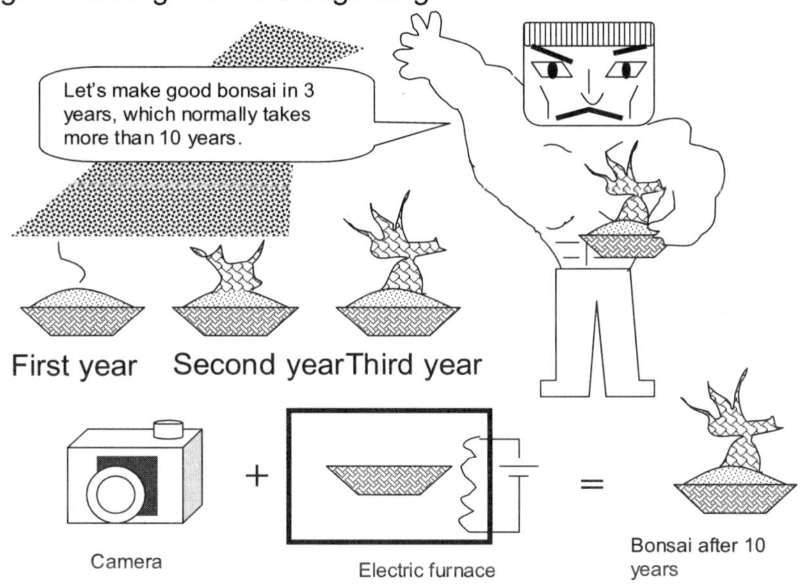

Fig.2 Karma and Forth rule

Booming solar cell business?

Booming solar cell business (Figure 1)

The solar cell business is booming.

Solar cells have long been developed. Why are they in the spotlight now?

On June 9, 2008, then Prime Minister Fukuda made a speech entitled *In pursuit of "Japan as a Low-carbon Society."* The summary is as follows.

1. Halving global CO_2 emissions by 2050: A reduction target that constitutes the core of the Cool Earth initiative.
2. Developing innovative technologies: Solar cells are one of major measures.

The solar cell market expanded first in Germany, where the government provided subsidies, and then in Italy and Spain. In these trends, solar cells have suddenly attracted attention as the third electronic technology after semiconductor and FPD technologies.

Is there any downside?

What are solar cells? (Figure 2)

Solar cells, or p-n junction diodes, have the same history as that of semiconductors. Taking it that the first practical solar cell was published in a paper by M. B. Prince of the Bell Telephone Laboratories in 1954, solar cells have a long history over 50 years. In Japan, the development of solar cells has accelerated since the 1974 oil shock. The recent rise of oil prices has fueled the solar cell business again.

Although it is advisable to consult related books for details, there are largely two types of solar cells: silicon-based and compound-based ones. Silicon-based solar cells are divided into bulk silicon solar cells cut from silicon ingots and thin film solar cells with silicon deposited on a glass substrate, as in FPD. Anyway, the cost is critical. Currently, the cost of power generation by solar cells is roughly 50 yen per KWh. It is desired that the cost will fall to 10 yen or less in the future. For solar cells, the power generation cost is a key factor.

The solar cell business is booming, and subsidies are provided; it may worth trying to develop solar cells.

Remember that some can take advantage of the boom, while some cannot.

Fig.1 Low CO_2 emission society by solar cell

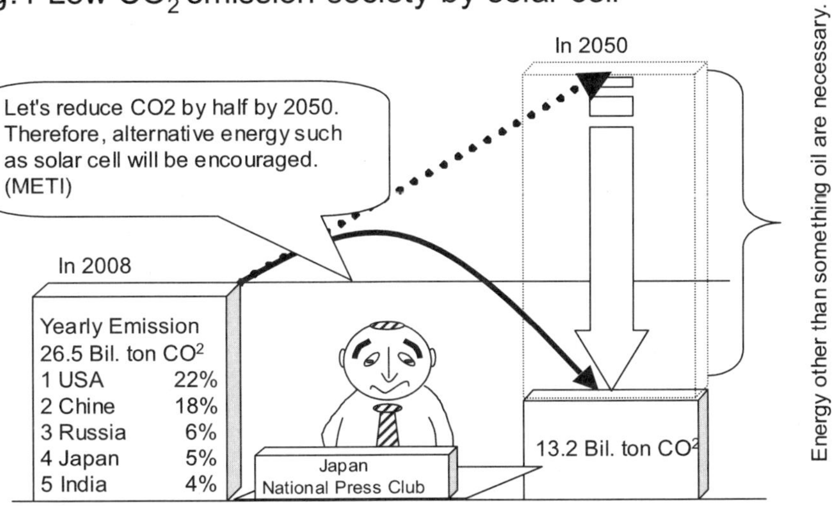

Fig.2 What are solar cells?

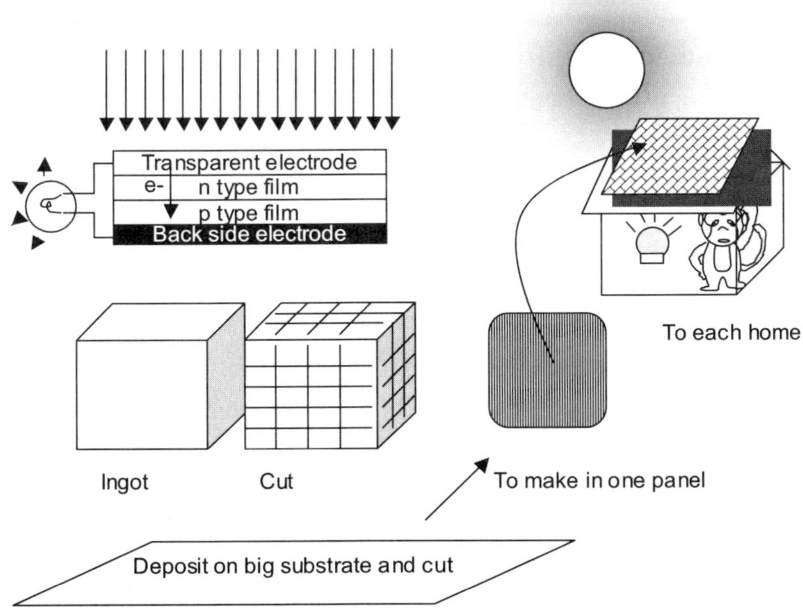

Looking at the forest of IC, FPD, and PV

It is recommended to look at the entire picture because development is like groping in the dark. Do not look only at the booming aspect of solar cells. Let's compare the trends of semiconductor devices (IC), flat panel displays (FPD), and solar cells (PV).

Comparison in size (Figure 1)

Let's start with comparison in size.

For semiconductor devices (ICs), the wafer size has increased from 150 mm to 200 mm, then to 300 mm, and is expected to be 450 mm in 2012.

For FPDs, substrates over 2000 mm are already in use. Compared to the FPD substrate size, a wafer size of 450 mm for ICs is much smaller.

The size of bulk silicon solar cells cut from silicon ingots is within 200 mm; the size of thin film solar cells is becoming as large as the FPD substrate size. There is a competition between bulk silicon solar cells, which have a size equivalent to the IC wafer size, and thin film solar cells, which are as thin as FPDs. Which will win depends on the power generation cost.

Looking in detail (Figure 2)

In Figure 2, which compares IC, FPD, and PV, the horizontal axis shows the thickness, and the vertical axis shows the substrate size. While the history of ICs is that of scaling and wafer size transitions, the history of FPDs is that of film thinning and substrate size increases. I dare to show the position of solar cells with reference to ICs and FPDs

It is said that the transition to 450 mm will be the final wafer size transition and that scaling to 16 - 11 nm will be the final scaling (although there are different opinions, I am not discussing them here).

FPDs only require the scaling of processing dimensions to several microns. However, their substrate size is approaching 3000 mm.

How about solar cells?

The processing dimensions are larger than those of FPDs. The substrate size of bulk silicon solar cells is smaller than the IC wafer size, and the substrate size of thin film solar cells is smaller than that of FPDs. There is room for improvement of solar cells. Here again, the power generation cost is critical. Development will be undertaken to reduce the power generation cost.

Fig.1 Comparison of substrate sizes

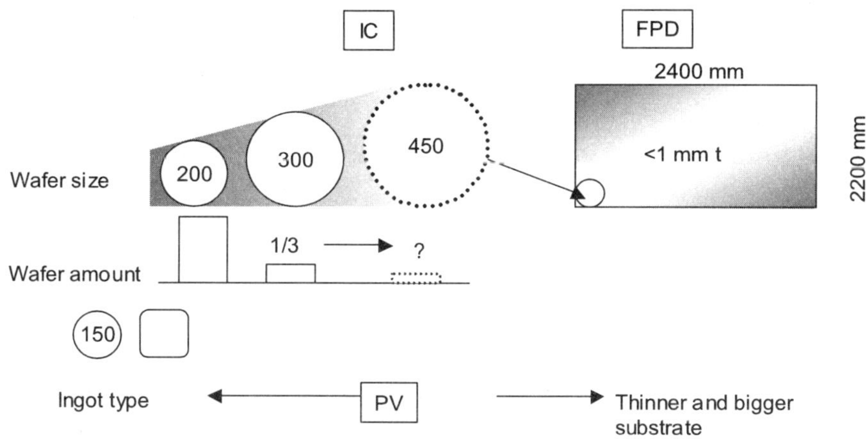

Fig.2 Correlation among IC, FPD and PV

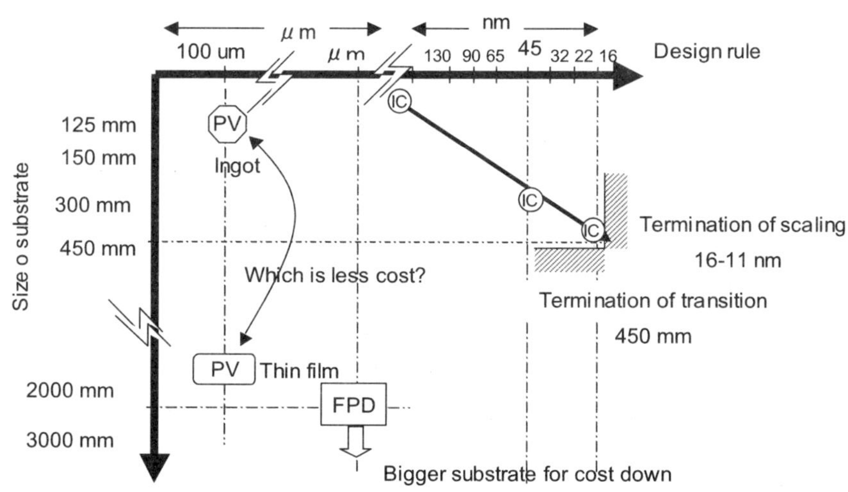

New York blackout!

Solar cells actually have downsides. To acquire new energy resources, various power generation methods, including solar cells, wind power, and fuel cells, have been developed and used for the last few decades. Do they have only benefits? Is there any downside?

New York blackout (Figure 1)

Lights in the hotel room started flickering at 4:10 pm on August 14, 2003. Shortly, the lights went out.
A major power failure hit New York.
Phones went dead, and water was cut off an hour later.
I was in New York then; I am a living witness of the history.
At 6 pm, the hotel was in a state of panic. It was still light outside, and travelers who still did not know what was going on were enjoying walking in the Central Park. Kumiko, my wife, walked back to the hotel from Chinatown. Her face was rigid, indicating how serious the situation was. Unlike me safe in the hotel, she was so nervous. The blackout occurred soon after she got off the subway. It could have been worse if she had been trapped in the train. It took three hours for her to walk back to the hotel. Imagine how hard it was to walk three hours thinking that I must be worried about her and that a riot might break out. Only those who experienced it know. In the InterContinental Hotel in front of the Central Park, power was restored at 4 am on August 15. It took another 12 hours before lights returned in Chinatown. Fortunately, no riot broke out, exhibiting that New York was well governed.

Caused by distributed generation (Figure 2)

It was later reported that the power outage was caused by an accident in a Canadian power plant, which led to chain blackouts in the power networks in the northeastern states. In Tokyo, the news of chain blackouts was received with surprise. In New York, many blamed it on the deregulation of the electric power industry.

If people sell electricity generated by solar cells, and its volume increases in the future, a similar problem may arise.

Tokyo Blackout in 20XX. What shall I do then...?

Fig.1 New York Blackout

Fig.2 Scary of distributed generation

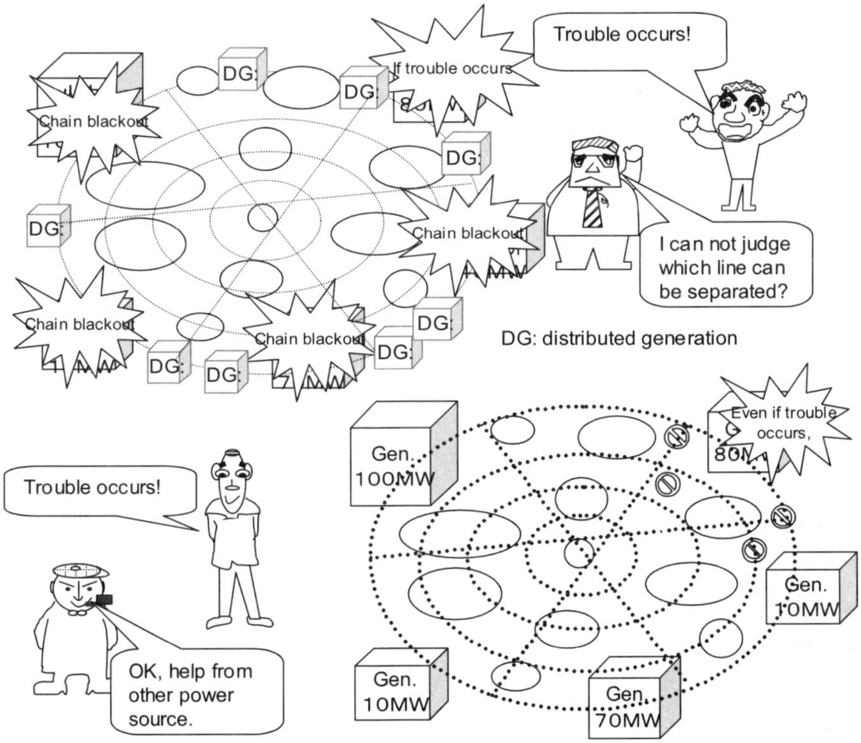

From Paradigm Shift 45 μ

Do you know TSV? TSV stands for Through Silicon Via. As well as PV described earlier, TSV was spotlighted as a promising technology in 2008. TSV and PV attract growing interest. What kind of technology is TSV? Let's look at TSV technology from the perspective of development.

From Paradigm Shift 45 (Figure 1)

As described in the section for Paradigm Shift 45, time has come for adopting 3D integration. There are obstacles to the scaling beyond 45 nm and wafer size transition to 45 cm in semiconductor device development. Thinner films for wafers (chips) are critical for 3D integration, but a film thickness of 45 μm and less is not easy to obtain.

Integration involves various technical terms you must learn. Figure 1 shows terms typically used in this field.

Conventionally, chips are mounted on a substrate; it's called SoB (system on board). Then, SoC (system on chip) integrating memory and logic into a device was introduced. Since some components cannot be readily integrated on a chip, SiP (system in package), a compromise solution to integrate memory and logic in a package, was developed (some may say that this is the perfect solution).

Figure 1 shows that the transition was made from SiP to SoC; in fact, however, the transition from SoC to SiP took place. Since then, a number of approaches to integration have been presented. If you come across terms like CoC (chip on chip) or PoP (package on package), just keep on reading. That is enough to start with. Anyway, there are too many technical terms for integration.

No restrictions on 3D integration? (Figure 2)

3D integration provides the three-dimensional stacking of chips on a single device, instead of conventional two-dimensional integration. 3D integration makes anything possible; you can try anything you want in a new field without restrictions. 3D integration may be easier to understand if it is described as playing with blocks.

TSV is the most remarkable technology in 3D integration. Why? The reason is described in the next section.

Fig.1 Paradigm is changed

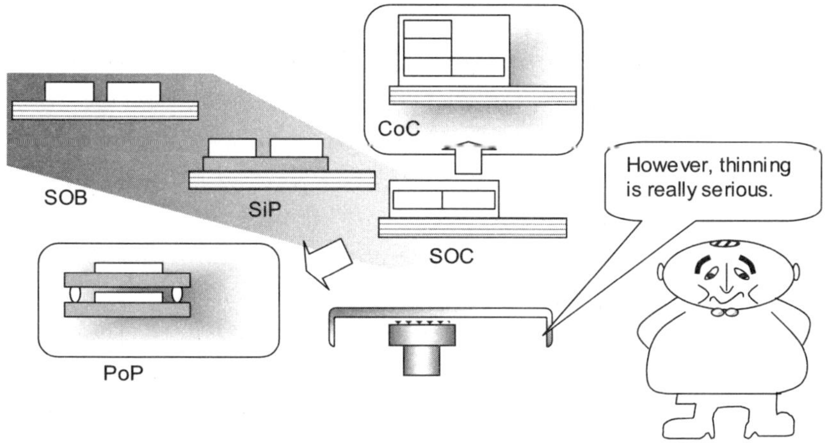

Fig.2 From 2 dimensions to 3 dimensions world

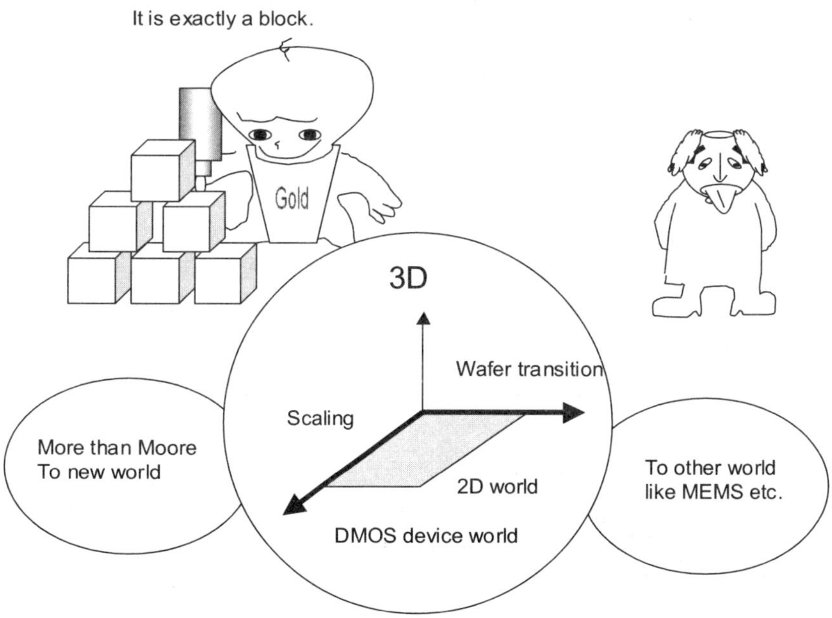

Based on plating technology

TSV technology using plating (Figure 1)

In this section, I would like to discuss TSV using an example of technology possessed by me (Ebara). As shown in Figure 1, there are largely four types of plating technology used for semiconductor devices.

Bump plating, which forms electrodes for devices, came first chronologically. Materials, including solder, gold, and copper, are used for plating.

Next, copper plating for multilayer interconnects was introduced. Multilayer interconnects are formed by the damascene process using copper, which has lower resistance than that of aluminum.

Further, cap plating for copper interconnect protection has been developed. For example, Co-W cap film is formed by electroless plating.

Finally, redistribution interconnect plating, instead of using conventional interposers, is applied to form redistribution interconnects in a device.

Plating for TSV filling has attracted attention as the fifth plating application in the future. Plating systems for TSV filling are similar to those used for bump and redistribution interconnect plating.

Is TSV technology a sprinter? (Figure 2)

TSV technology is currently applied only to CMOS image sensors for sensor miniaturization. The application of TSV technology aims at (1) miniaturization, (2) enhancement, and (3) diversification. Remember that the cost is always critical.

Compared with wire bonding technology, (1) miniaturization is apparently achieved. In addition, a much shorter interconnect length allows fabricating high-speed devices.

Indeed, TSV technology is like a slim (miniaturized) fast sprinter.

Plating for TSV filling (Figure 3)

For reference, the images of TSV filled by plating are shown in Figure 3. A 20 μm via gradually filled from the bottom is impressive, isn't it?

This is TSV technology.

Fig.1 From several plating application

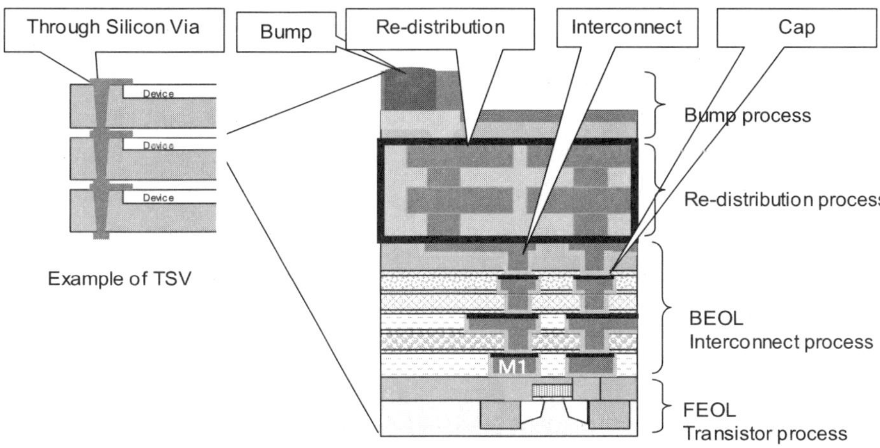

Fig. 2 Purpose of TSV

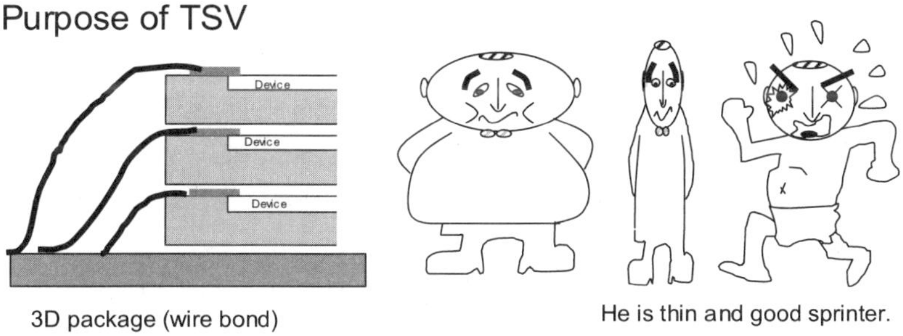

Fig.3 Plating image

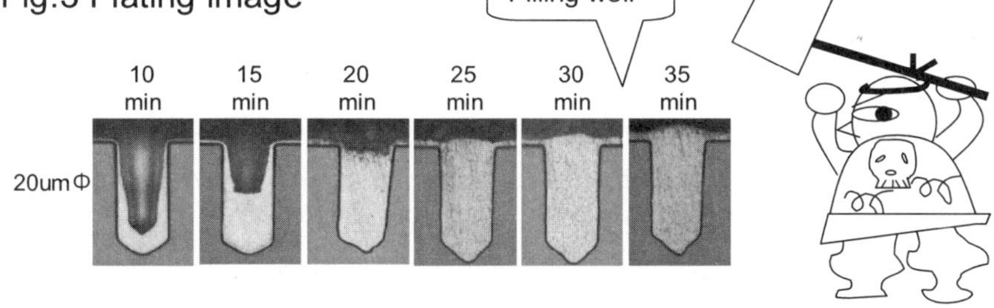

Booming MEMS technology

Definition of MEMS (Figure 1)

MEMS stand for microelectromechanical systems. They were variously referred to as micromachines, micro system technology, etc., but now are generally defined as MEMS. Furthermore, such systems at the nano level are particularly called NEMS (nanoelectromechanical systems).

MEMS are applied in diverse fields, including semiconductor, mechatronics, chemical plants, and biotechnology. The MEMS technology is mainly applied to semiconductor processing. The purpose of the application is *miniaturization*, *downsizing*, *scaling*, and *minimization*, of course. The size range is also wide: mm, μm, and nm. For reference, the relationship between the dimensions of MEMS and the fields of application, from DNA chips to chemical plants, is shown in Figure 1.

Overview of the evolution of MEMS (Figure 2)

There are three streams of the development of MEMS.

The first one is the evolution from silicon technologies. SiP (logic/memory chips and other components of semiconductor devices are enclosed in a package) or SoC (those are integrated into a chip) evolved into mechatronics (MEMS of today. Even sensors and actuators are integrated). MEMS is now applied to a wider fields: a μTAS, gene chips, acceleration sensors, ink-jet, etc. These are some recent examples of the evolution of MEMS based on silicon technologies.

The second one is the evolution from metal (mechanical) technologies. A micro reactor is indeed a chemical plant at the mm or μ level downsized using the MEMS technology. The micro reactor does not fall into mechatronics. That may be because a combination with ICs is not specified.

The last one is the evolution from the field of organic chemistry. Organic chemical manufacturers started the development of functionalization and miniaturization technologies based on bottom-up technologies.

Note) Top-down technologies are defined as those used to form interconnects by top-down approaches, such as deposition and removing deposited material. In contrast to this, bottom-up technologies are those used to form interconnects from the bottom of trenches by using carbon nanotubes, etc.

Fig.1 Application with scale

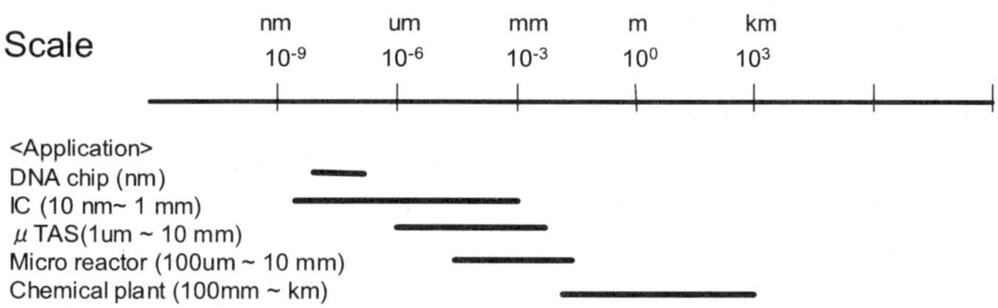

Fig.2 Three flows of evolution

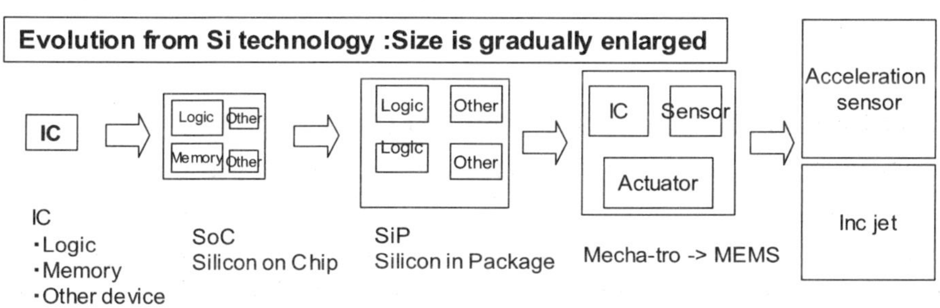

Evolution from metal technology : Size is gradually reduced.

Chemical plant, pumps etc. -> Micro reactor -> μTAS

Evolution from Organic chemistry :

The Bottoms improvement technology etc. appear.

DNA chip based on MEMS technology

What is a DNA chip?

A DNA chip is also referred to as a biochip or a gene chip. Terms, such as a μTAS (total analysis system) and a BioMEMS, are also used as the one that have similar meaning to the DNA chip.

Composition of the human body (Figure 1)

It is said that the human body is made up of about 60 trillion cells. The current world population is about 6 billion. Each cell contains a nucleus and chromosomes. A chromosome, in a form of a very long chain, is a chemical substance, DNA (deoxyribonucleic acid), which is composed of phosphoric acid, sugar, and bases. The bases play important roles in genes. Four bases are found: (A) Adenine, (T) Thymine, (G) Guanine, and (C) Cytosine.

As you may have heard from recent news that DNA distinguishes an individual from the others, it is the four bases that play key roles in the identification of individuals. Sequences of the four bases tell us all the information about human beings. The DNA chip is used to analyze the sequences.

DNA chips and semiconductor plants (Figure 2)

Various types of the DNA chip are available. Some types are fabricated using semiconductor technologies. Three billion genes need to be analyzed. This number is equivalent to 3 gigabytes. There are no technology but semiconductor technologies that can handle such a volume of data.

The four bases need to be well arranged on a DNA chip. The arrangement is just the same as the semiconductor technologies, such as lithography, deposition, removing deposited material, and cleaning.

Figure 2 shows similarities between a semiconductor plant and a DNA chip plant. I have heard that one DNA chip manufacturer purchased a 6-inch semiconductor wafer plant and converted it to a DNA chip plant. The company must have purchased the 6-inch wafer plant at a bargain price since 12-inch wafers are dominant in today's semiconductor industry.

What do you think about the story above?
This is an example of the MEMS technology applied to the biotechnology.

Fig.1 DNA ,genome and gene

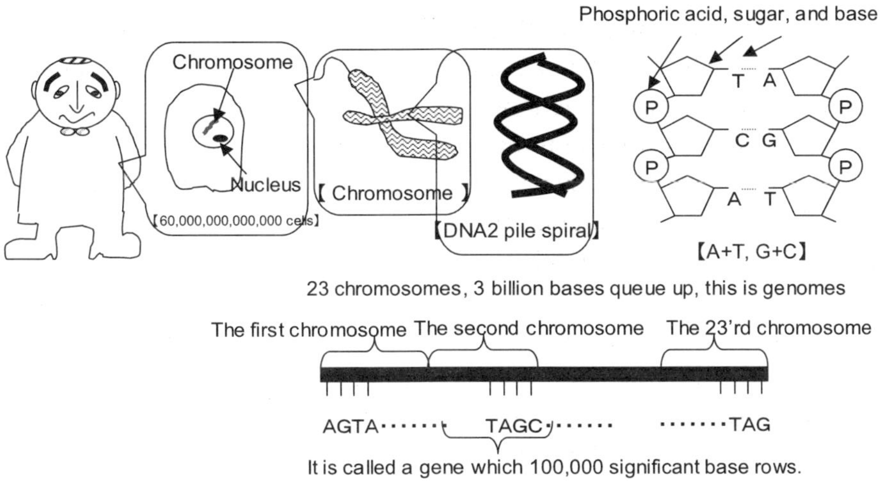

Fig.2 Similarity of semiconductor factory and DNA chip factory

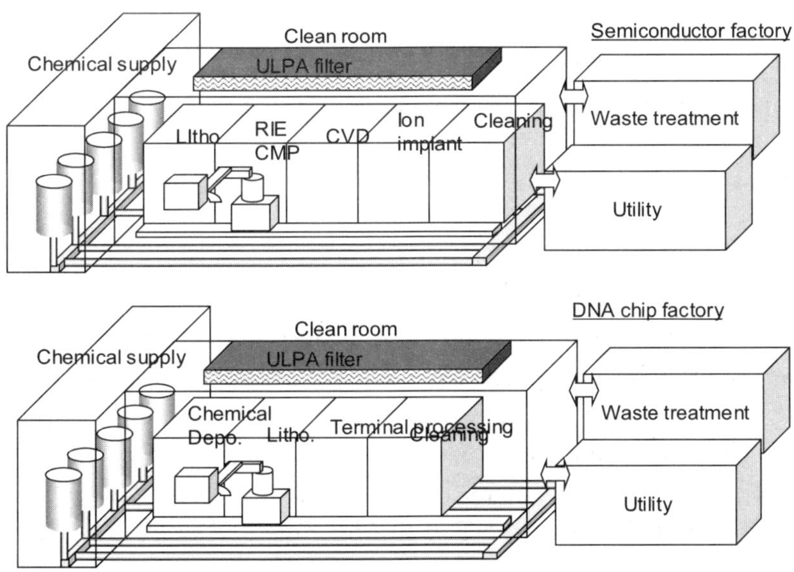

Things you cannot predict in development

In an information society (Figure 1)

How the world will change in an Internet or information society?

I have a right and obligation to consider the future of the world as a person who has been involved in the semiconductor field. I present my consideration below whether the future is bright or dark or a paradise or a hell.

The first question is what an information society is like.

Broadband? Web 2.0? Such terms are all Greek to the general public.

"I heard that terrestrial digital broadcasting means my TV set will show no picture any longer," said my mother. That's true in a sense...

Briefly speaking, the *volume* and the *speed* of information have been magnified more than ever before. In the future, we will find it increasingly difficult to handle information, which used to be manageable.

I wonder if we, engineers in the semiconductor field, work for making the world chaotic and full of fickle people at a mercy of a storm of information.

Recommended movies - IT, BT, and NT (Figure 2)

Some movies envision our future. Surprisingly, some events in such movies have turned into reality. My recommended movies are: (1) *Astro Boy*, (2) *Fantastic Voyage*, (3) *The Island*, (4) *The Matrix*, and (5) *Live Free or Die Hard* (*Die Hard 4.0*).

(1) (IT) We are already in the robot age though real robots are not as sophisticated as Astro Boy.

(2) (IT, BT, and NT) In *Fantastic Voyage*, human beings were injected into a human body. Now a MEMS-based endoscope is inserted into a human body.

(3) (BT) In *The Island*, human beings were cultivated. Cultivating human beings is not yet possible but regenerative medicine has advanced so much.

(4) (IT) The world depicted in *The Matrix* series (especially, the first film) will turn into reality soon by Linden Lab that runs Second Life.

(5) Finally, if cyber terrorism occurs in such an advanced information society based on the fusion of IT, BT, and NT as in *Live Free or Die Hard*, what would happen?

Fig.1 Information overflows.

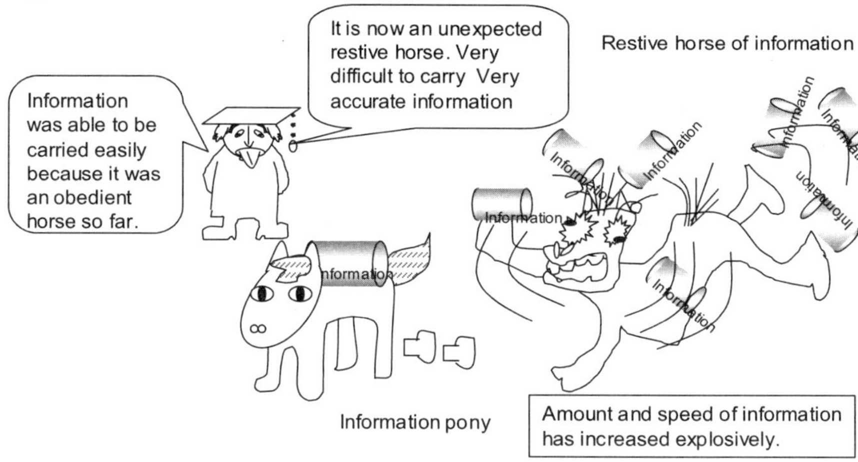

Fig.2 Let's see the movie.

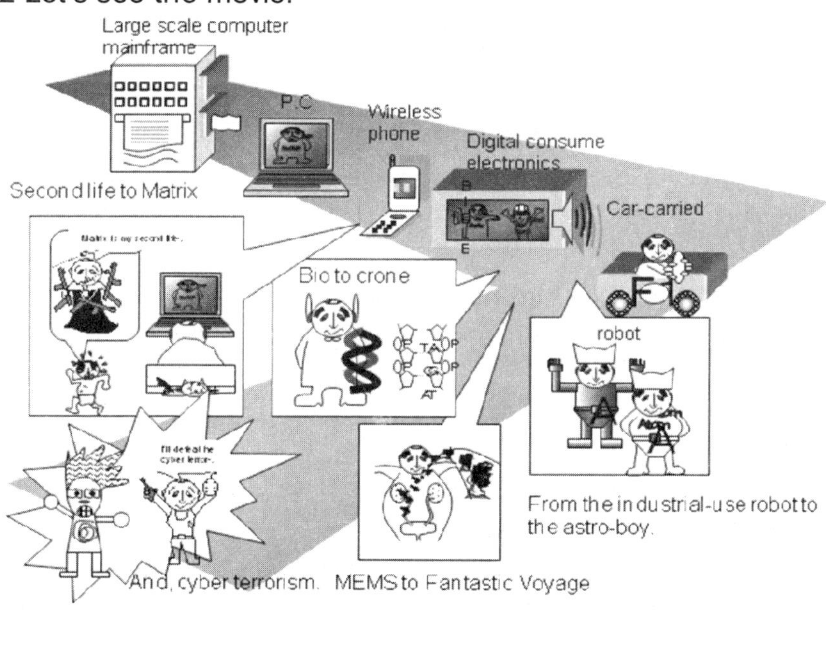

Save the Earth

Five Rules (Figure 1)

Development is like groping in the dark. Follow the doctrines and five rules.

JOY - Defy Common Wisdom (Figure 2)

Defy common wisdom. But what is common wisdom? Say *JOY* in a loud voice. JOY (Defy common wisdom)!

How to Survive Using R&D Concepts (Figure 3)

Mr. Yamada, newcomer, now knows what common wisdom is and how to defy it. If all departments of a company optimize their capabilities for development, you and your company will be greatly benefited.

R&D Born through Education (Figure 4)

Development may be born through education. In-house education and external study opportunities will help you. If you are a development engineer, I advise you to get a PhD. It will change your world.

How to Write Papers and Patents (Figure 5)

Write patents for your company. Write papers for society. Contribute to your company first, and then to society. This is for your own benefit.

Five Revolutionary Technologies (Figure 6)

Necessity is the mother of invention, and desire is the father of revolutionary technologies. The five gods of development have shown you the examples of innovative technologies. I hope that you find them useful.

Let's Look at the Forest of Technology Development

We come to the end of this book.

At age 57, I don't think that I am experienced enough, but I dare to publish this book since some people in universities and Ebara found it useful. Although I cannot usually see the forest for the trees, I was able to "look at the forest of technology development" through this book. I hope that the book will also help you. Thank you for reading.

Fig.1 Five rules

1. Pursue Something New.
2. Develop *must* technologies, not better technologies.
3. Think development as a cost that must be recovered.
4. Start with possessed technologies (use your expertise).
5. Think and think! Think until you feel sick!

Fig.2 **JOY - Defy Common Wisdom**

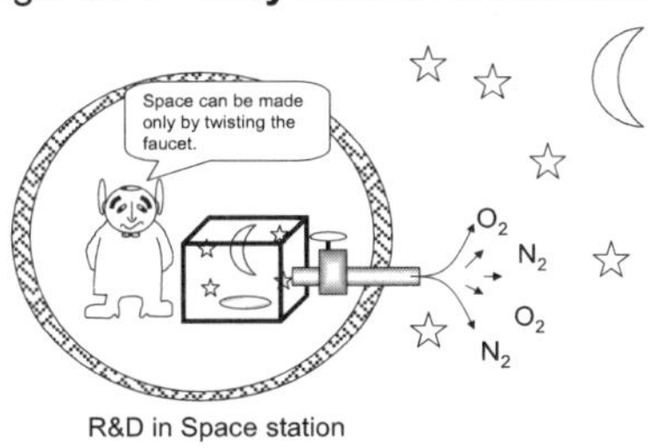

Fig.3 How to Survive Using R&D Concepts

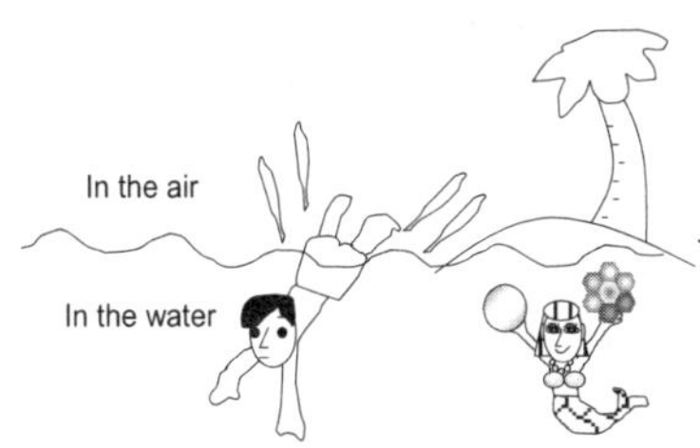

Fig.4 R&D Born through Education

Fig.5 How to Write Papers and Patents

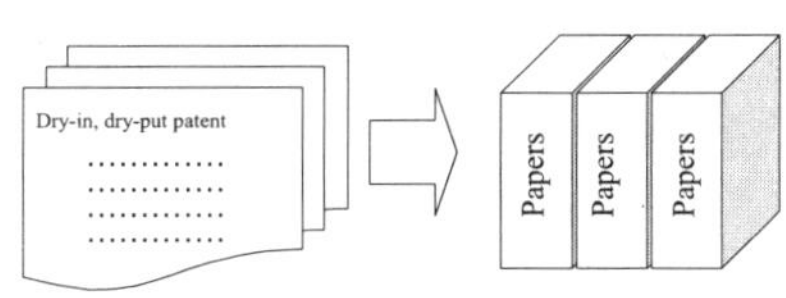

Fig.6 Five Revolutionary Technologies

From an inhabitant of the forest of technology development

Do you want to know who I am?
Put aside the question. What are humans doing?
The other day, a human came to me and bought heavy black smoke spreading poison (dioxin).
If such black smoke is useful, they should produce more, but some are trying to reduce it. Human behavior is beyond my understanding.
What? Am I wrong?
He came to buy CO_2 emission rights as a measure against global warming?
That makes me more confused.
In the old days, they used to exchange goods by barter.
Then, they started exchanging rice with paper sheets (paper money).
Later, they started to use hard plates (credit cards) to buy not only rice but also anything they want.
Now, anything can be purchased through humming electric waves (Internet) traveling around the Earth.
Their population is 6.5 billion now and will be 9 billion soon.
The Earth is not rich enough (in food and resources) to feed such a large number of people.
It seems that there are two types of humans.
Those who are eager to earn money and those who are weak but trying to save the Earth.
I have sent my disciples to the latter. Five gods, give engineers a help!
If nothing is done, I (Earth) will die. Do something while I am still alive...

This is a dream I had.
Our individual capabilities may be small, but I am sure that we can save the dying Earth by combining our development capabilities. Let's combine the development capabilities of scientists and engineers, and develop technologies to save the Earth. I will make my humble contributions. Let's do our best!

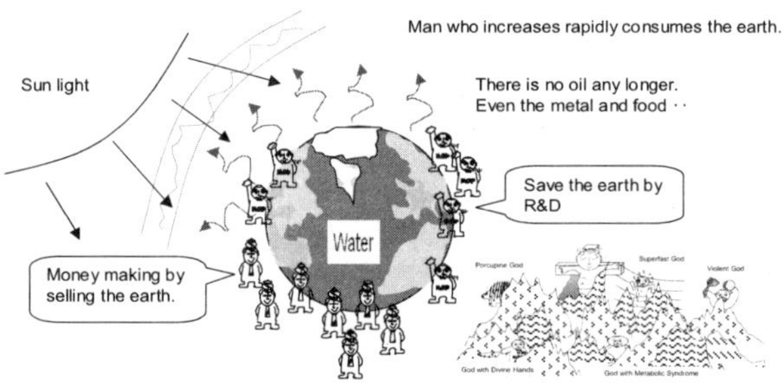

References

1 K. Numagami: ‘Organization Design’, Nikkei Bunko
2 M. Tsujimura: ‘Semiconductor Equipment for new materials’, Denshi Zairyou, May(2001), p28-32
3 M. E. Porter: ‘Competitive Strategy’, Diamond Co.
4 Henry Chesbrough: ‘Open Innovation’, Sangyo Noritsu University
5 M. Tsujimura : ‘Defect Free Cleaning in the 65 nm TN beyond’, Semicon Korea 2005 Proceedings, 2005
6 M. Tsujimura : ‘Clean CMP System’, Clean Technology (1995),p31-33
7 ITRS(International Technology Roadmap for Semiconductors), SIA, (2007)
8 M. Tsujimura et al.: ‘Project management of semiconductor wet process development’, Project Management, Feb.(2004)
9 M. Tsujimura : ‘Electro-plating tool for Cu interconnect’, Denshi Zairyo, Mar.(2000), p45-50
10 M. Tsujimura et al: ‘General Principle of Planarization Governing CMP, ECP, ECMP & CE’,2004 VMIC Conferences, Proceedings 2004 IMIC-030/00/0267, p267-274
11 M. Tsujimura et al: US5885138, Method and Apparatus for DRY-IN DRY-OUT Polishing and washing of a semiconductor device
12 K. Toyoda: ‘Kaisha-Ho’. Natsume Co.
13 JSME Journal, Vol. 111, No.1070, Jan.(2008)
14 http://www.clarkson.edu/camp/
15 Y. Mochizuki, M. Tsujimura et al: ‘FEM Aalysis of the maximum stress in vias with the damascene structure’, JSME proceedings, Mar.(2005)
16 M. Tsujimura et al: ‘Paradigm Shift 45 – Drastic changes in semiconductor field’, Semiconductor News Paper Jan.(2006)
17 M. Tsujimura et al: ‘Paradigm Shift 45 – Ver.2.0’, Semiconductor News Paper Jan.(2007)
18 I .Noji, M. Tsujimura et al: ‘Application of Electro-Chemical Polishing in DI water to Cu damascene wiring planarization process’, MRS Advanced Metallization Conference 2001 （AMC2004 , Proceedings）,(2004),
19 Y. Ishi, M. Tsujimura et al: ‘Upgrade of polish profile by magnetic control head’, JSME proceedings, Oct.(2003), p281
20 M. Tsujimura et al: ‘A Novel Compact ECD Tool for ULSI Cu Metallization’, ISSM2000, Sponsored by UCS,IEEE EDS and SEMI ,September 26-28,2000, p106-109
21 M. Tsujimura et al: ‘Development of semi-conductor device cleaning by cavitation jet’, JSME journal, 67-653, (2001), p88-94

Index